Springer Theses

Recognizing Outstanding Ph.D. Research

Aims and Scope

The series "Springer Theses" brings together a selection of the very best Ph.D. theses from around the world and across the physical sciences. Nominated and endorsed by two recognized specialists, each published volume has been selected for its scientific excellence and the high impact of its contents for the pertinent field of research. For greater accessibility to non-specialists, the published versions include an extended introduction, as well as a foreword by the student's supervisor explaining the special relevance of the work for the field. As a whole, the series will provide a valuable resource both for newcomers to the research fields described, and for other scientists seeking detailed background information on special questions. Finally, it provides an accredited documentation of the valuable contributions made by today's younger generation of scientists.

Theses may be nominated for publication in this series by heads of department at internationally leading universities or institutes and should fulfill all of the following criteria

- They must be written in good English.
- The topic should fall within the confines of Chemistry, Physics, Earth Sciences, Engineering and related interdisciplinary fields such as Materials, Nanoscience, Chemical Engineering, Complex Systems and Biophysics.
- The work reported in the thesis must represent a significant scientific advance.
- If the thesis includes previously published material, permission to reproduce this must be gained from the respective copyright holder (a maximum 30% of the thesis should be a verbatim reproduction from the author's previous publications).
- They must have been examined and passed during the 12 months prior to nomination.
- Each thesis should include a foreword by the supervisor outlining the significance of its content.
- The theses should have a clearly defined structure including an introduction accessible to new PhD students and scientists not expert in the relevant field.

Indexed by zbMATH.

Yui Tik (Andrew) Pang

Biomolecules' Conformational Changes Studied by Simulations and Enhanced Sampling

Doctoral Thesis accepted by Georgia Institute of Technology, USA

Springer

Yui Tik (Andrew) Pang
School of Physics
Georgia Institute of Technology
Atlanta, GA, USA

ISSN 2190-5053 ISSN 2190-5061 (electronic)
Springer Theses
ISBN 978-3-031-70601-1 ISBN 978-3-031-70602-8 (eBook)
https://doi.org/10.1007/978-3-031-70602-8

This Springer imprint is published by the registered company Springer Nature Switzerland AG
The registered company address is: Gewerbestrasse 11, 6330 Cham, Switzerland

For every teacher and mentor that helped me get here.

Supervisor's Foreword

Biomolecules are responsible for life itself, yet they are subject to the same physical laws as inanimate matter. In characterizing Brownian motion in 1827, Robert Brown observed that pollen grains were moved by the same seemingly random forces as glass fragments and stone dust, but it took nearly a century for Albert Einstein to lay down the mathematical underpinnings describing this motion (1905). Decades later, in his Lectures on Physics, Richard Feynman famously said "everything that living things do can be understood in terms of the jigglings and wigglings of atoms." This profound idea, itself the culmination of centuries of observation, spurred the development of techniques for modeling the behavior of biomolecules through the rigorous application of physical laws, collectively called "molecular dynamics." Since the first simulation of a protein by Andrew McCammon, Bruce Gelin, and Martin Karplus in 1977 [1], it has been increasingly recognized that biomolecules are dynamic entities, undergoing both local and global conformational changes. These changes must be sampled in order to fully describe the function of biomolecules within the living cell.

In his thesis, Dr. Andrew Pang takes on the challenge of sampling biomolecular motions with aplomb. Starting with fundamental methods, he tackles the ever-present problem of determining accurate force-field parameters for small molecules, which bind to and modulate protein conformations and function. Specifically, he develops new methods for developing parameters incorporating the so-called "σ-hole," which represents an anisotropic charge distribution around some halogen atoms. He applies this to the small molecule AT-130, a potential drug against Hepatitis B Virus. In the next chapter, he scales up to the folding of the secreted domain of the autotransporter pertactin. He uses multiple simulation approaches, including steered molecular dynamics and free-energy calculations, to investigate potential folding pathways of this >1000-residue protein. His results demonstrate how the "vectorial" folding pathway, in which the domain folds as it is secreted across the bacterial membrane, is more favorable than the isolated folding pathway. Finally, he takes on the largest system yet: the full spike protein trimer of SARS-CoV-2, the virus responsible for the recent worldwide pandemic. In this study, he uses advanced replica-exchange umbrella sampling simulations on one of the fastest supercomputers in the world at that time, Summit at Oak Ridge National Laboratory. These simulations, 160 μs in total, provided unparalleled insights into the dynamics of spike-protein opening and how it is influenced by the numerous glycans that are attached to the protein's surface. In total, the work presented by Dr. Pang here not only makes a notable impact on our understanding of the structure-dynamics-function relationship of each respective subject, it also provides a roadmap to the application of advanced simulation methods for other systems of interest, now and well into the future.

Dunn Family Professor of Physics
Georgia Institute of Technology
Atlanta, GA, USA

James C. Gumbart

Reference

1. J. A. McCammon, B. R. Gelin, and M. Karplus, "Dynamics of folded proteins," *Nature*, vol. 267, no. 5612, pp. 585–590, 1977.

Acknowledgments

I would like to thank my advisor, J.C. Gumbart, for his help in preparation of this work. I would also like to thank the members of my thesis committee for their valuable feedback and advice to alleviate this thesis. Special thanks to my family, friends, and colleagues for the inspiring discussion and supports throughout my academic journey.

The author gratefully acknowledges the support for part of this work offered by a Texas Advanced Computing Center (TACC) Frontera Fellowship. Frontera is supported by NSF grant no. OAC-1818253. Any views and conclusions contained herein are those of the author, and do not necessarily represent the official positions, express or implied, of the funders.

Parts of this thesis have been published in the following journal articles:

1. **Pang, Y. T.**; Pavlova, A.; Tajkhorshid, E.; Gumbart, J. C. Parameterization of a drug molecule with a halogen σ-hole particle using ffTK: Implementation, testing, and comparison. *J Chem Phys* **2020**, *153* (16), 164104. AIP Publishing.
2. **Pang, Y. T.**; Acharya, A.; Lynch, D. L.; Pavlova, A.; Gumbart, J. C. SARS-CoV-2 spike opening dynamics and energetics reveal the individual roles of glycans and their collective impact. *Commun Biol* **2022**, *5* (1), 1170. Nature Publishing Group UK London.
3. **Pang, Y. T.**; Hazel, A. J.; Gumbart, J. C. Uncovering the folding mechanism of pertactin: A comparative study of isolated and vectorial folding. *Biophys J* **2023**, *122* (14), 2988–2995. Elsevier.

Acknowledgments

Contents

List of Acronyms

AbASA	antibody accessible surface area
ACE2	angiotensin-converting enzyme 2
BAM	β-barrel assembly machinery
CAM	capsid-assembly modulator
CEST	chemical exchange saturation transfer
CGenFF	CHARMM General Force Field
cryo-EM	Cryo-electron microscopy
FAST	fluctuation amplification of specific traits
ffTK	Force Field Toolkit
FRET	fluorescence resonance energy transfer
HBV	hepatitis B virus
HMR	hydrogen mass repartitioning
MBAR	Multistate Bennett Acceptance Ratio
MD	molecular dynamics
MEP	minimum energy path
MFPT	mean first passage time
ML	machine learning
MM	molecular mechanical
MSM	Markov state model
NMR	nuclear magnetic resonance
NTD	N-terminal domain
OM	outer membrane
ORNL	Oak Ridge National Lab
PDB	Protein Data Bank
PES	potential energy surface
PME	particle-mesh Ewald
PMF	potential of mean force
ProGAN	progressive growing generative adversarial network

proK	Proteinase K
QM	quantum mechanical
RBD	receptor binding domain
RBM	receptor binding motif
REUS	replica exchange umbrella sampling
RMSD	root mean square deviation
S	spike
SASA	solvent accessible surface area
SLUS	self-learning adaptive umbrella sampling
SMD	steered molecular dynamics
SPIB	state predictive information bottleneck
US	umbrella sampling
VACF	velocity autocorrelation function
VAE	variational autoencoder
VMD	Visual Molecular Dynamics
WHAM	weighted histogram analysis method
WT	wild-type

Chapter 1
Introduction and Background

1.1 Biomolecules Flexibilities

Within every living organism, myriad biomolecules constantly interact and undergo reactions to sustain essential cellular functions [1]. These biomolecules vary widely, both in size and function. At the macro level, we have macromolecules: proteins that catalyze various biochemical reactions and provide structural integrity to tissues, carbohydrates that serve as primary energy sources and offer structural support, lipids that form cellular membranes and store energy, and nucleic acids that carry genetic information and regulate protein synthesis. On the smaller scale, there are molecules like vitamins that support enzymatic reactions, and hormones that transmit signals and modulate protein activities. Each biomolecule plays its specific role, and collectively, they ensure the organism remains in good health and enable it to overcome the diverse challenges presented in Nature.

Biomolecular flexibility is intrinsic to their functionality, rooted deeply in their molecular architecture. This adaptability arises predominantly from the capacity of these molecules to undergo conformational alterations, either as a response to binding events or shifts in environmental conditions [2, 3]. Central to this flexibility are rotatable chemical bonds, essentially those constituted solely by a σ-bond. These bonds possess the ability to rotate, thereby reshaping the broader conformation of the molecule. The energy landscape governing such rotations is influenced by both the intrinsic chemical attributes of the bond and its ambient chemical milieu. Such elements dictate the bond's preferred torsion angles and its torsional resilience. Integrating these factors with other inter- and intra-molecular interactions, a few global conformations often stand out as energetically favorable states [4, 5].

Y. T. (Andrew) Pang, *Biomolecules' Conformational Changes Studied by Simulations and Enhanced Sampling*, Springer Theses,
https://doi.org/10.1007/978-3-031-70602-8_1

1.1.1 Small Molecules

In the realm of structural biology, small molecules are characterized as biological compounds with relatively low molecular weight, typically falling below the 1 kDa threshold [6]. While many of these molecules serve foundational roles, acting as precursors or building blocks for their more complex macromolecular counterparts—amino acids for proteins, saccharides for carbohydrates, nucleotides for nucleic acids, and so forth—others play pivotal roles as essential metabolites, orchestrating a vast array of biological processes from cellular signaling to defense mechanisms. Beyond these naturally occurring entities, many drugs are also categorized as small molecules. These pharmaceutical agents are specifically formulated to target and modulate protein functions, either by inhibiting their inherent activities or by disrupting their interactions with partner proteins [7].

In drug design, accurately accounting for the accessible conformations of molecules under physiological conditions is pivotal for success. Many molecules possess rotatable bonds, enabling them to undergo conformational changes due to thermal fluctuations or in response to their environment—be it in a solvent or when bound to a protein. While the flexibility of drug molecules can be advantageous by facilitating optimal docking poses with target proteins, it might also introduce a significant entropic cost upon binding [8]. From a statistical physics viewpoint, these considerations are vital when determining solvation energy [9], lipophilicity energy [10], and perhaps most crucially, the binding free energy with the target macromolecule [11]. Thus, understanding the dynamism of small molecules is a cornerstone in drug design, setting the stage for the even more intricate dance of flexibility observed in larger entities, like proteins.

1.1.2 Proteins

Central to numerous bioactivities, proteins execute a vast array of cellular functions, from enzymatic catalysis to gene regulation. Given their pivotal roles, they often emerge as primary drug targets, either to modulate aberrant proteins in human diseases or to thwart proteins in pathogens. Remarkably, the diversity of protein functions is orchestrated using only 20 primary amino acids. Distinguished by their unique side chains, these amino acids are typically grouped into four categories: charged, polar, hydrophobic, and special. Covalently linking together, these amino acids create a linear peptide chain, often referred to as the backbone. Both the backbone and the side chains harbor numerous rotatable bonds, paving the way for a vast conformational landscape. Yet, under physiological conditions, proteins adopt specific secondary and tertiary configurations known as their native structures. These native configurations are paramount for protein functionality, often featuring specialized regions, such as active sites for biomolecular interactions or channels for selective substance transport. Misfolded proteins jeopardize cellular functions and are either salvaged by chaperone proteins or targeted for proteasomal degradation [12]. Grasping protein folding is indispensable for elucidating their functional mechanisms and for deciphering protein aggregation—a key feature of neurodegenerative ailments like Alzheimer's [13].

Proteins' dynamism is not confined to their primary folding process. Analogous to how a car engine undergoes cycles to mobilize the wheels, proteins frequently oscillate between various conformations to fulfill their roles [14–17]. These dynamics span a spectrum—from subtle side chain rearrangements of amino acids to extensive shifts involving protein domains or subunits. Certain proteins are enriched with intrinsically disordered regions, conferring flexibility and facilitating diverse conformations contingent on their interacting partners [18, 19]. Deciphering the intricate interplay between protein structure, dynamics, and function is paramount for unwrapping the molecular mechanisms steering biological events, and by extension, strategizing therapeutic interventions.

1.2 Role of Computational Simulations

The imperative of understanding biomolecular structural dynamics has driven extensive research in the field. Over the past several decades, remarkable strides have been made in both experimental and computational techniques to elucidate biomolecular dynamics. These advancements include the evolution of chemical exchange saturation transfer (CEST) nuclear magnetic resonance (NMR) spectroscopy [20] and the pioneering of dedicated molecular dynamics (MD) simulation platforms like Anton [21]. Computational simulation methods offer distinctive advantages, chiefly the ability to exert precise control over—and measure—every facet of the simulated environment across varied scales. This level of control, combined with its cost-effectiveness relative to wet-lab experiments, positions simulations as a powerful research tool. Among the array of simulation techniques, all-atom MD simulations stand out for biomolecular studies. Their appeal lies in the optimal trade-off they provide between computational speed and detail. In these simulations, every component of the biological system—from the biomolecules of interest to their surrounding environment—is represented explicitly at an atomistic level. Using modern computational resources, all-atom MD simulations can achieve speeds approximating 160 ns/day for standard system sizes encompassing $\sim$1000 nm^3 and $\sim$100K atoms [22].

Nevertheless, many critical biological phenomena, such as ligand binding, protein folding, and large-scale conformational changes, occur on millisecond timescales or even longer. This temporal challenge prompts many researchers to explore avenues for accelerating MD simulations. One prominent category of these strategies is termed "enhanced sampling". The family of enhanced sampling is broad and actively growing, ranging from simply applying a force to pull the atoms of the biomolecule from one reference structure to another [23], to sophisticated methods such as selectively heating up part of the system to speed up conformation sampling [24]. In general, however, the central philosophy of enhanced sampling hinges on promoting the observation of rare events during the simulation. Many enhanced sampling also include a mechanism to rectify any introduced biases in post-analysis based on principles from statistical mechanics, enabling the accurate calculation of key metrics like free energy changes.

1.3 Biomolecular Systems Under Investigation

In this thesis, I delve into the intricate dynamics of three distinct biomolecules, utilizing both MD and enhanced sampling techniques. While these biomolecular systems vary in size and nature, they collectively highlight the power and versatility of simulations and enhanced sampling in elucidating the complexities of biological phenomena. This thesis aims to offer a comprehensive account of the challenges and successes in decoding the dynamical behaviors of these biomolecules.

1.3.1 Capsid Assembly Modulator AT130

Our exploration begins with AT130, a therapeutic small molecule known for its role in promoting premature assembly of the hepatitis B virus (HBV) nucleocapsid, leading to the formation of empty capsids [25]. A pivotal feature of AT130 is the flexible torsion at the center of the molecule, allowing it to adopt three different metastable conformational states [26]. However, due to the diversity of small molecules, accurate force field parameters specific to the molecule under study must first be determined to accurately simulate its dynamics in MD simulations. These parameters, while obtainable by CHARMM General Force Field (CGenFF) by analogy, may not accurately capture the interactions, especially for a molecule with a sensitive torsion like AT130. We optimized the force field parameters using Force Field Toolkit (ffTK), fitting the molecular mechanical energies to the quantum mechanical (QM) energy landscape. We demonstrated the parameter fitting process, particularly the use of a new virtual particle introduced in CGenFF 4.0, which represents the small positive charged region on halogens [27]. Using the parameters from CGenFF analogy before and after ffTK fitting, we ran MD simulations of AT130 in water and in its target protein binding pocket to compare the effectiveness of ffTK parameter fitting and the usage of the new virtual particle [26].

1.3.2 Passenger Domain of Pertactin

Transitioning to proteins, we studied the folding dynamics of the pertactin passenger domain from *Bordetella pertussis*, providing insights into the secretion processes of virulence factors in Gram-negative bacteria. Previous studies have shown that when autotransporters travel from the inner to outer membrane through the periplasm, they remain unfolded until they reach the outer membrane. The prevailing hypothesis suggest that folding is unfavorable in isolation and only becomes favorable when the autotransporter is guided through the secretion barrel at outer membrane, folding sequentially from the C- to N-terminus [28]. We tested this hypothesis using steered molecular dynamics (SMD) and self-learning adaptive umbrella sampling (SLUS) simulations, and conducted a detailed investigation of the hydrogen bonds and other residue-residue interactions that occur during the folding process. This analysis provided molecular insights into the forces at play, revealing key interactions that drive the folding process and influence the final protein structure [29].

1.3.3 SARS-CoV-2 Spike Protein

The most extensive biomolecule examined in this thesis is the trimeric SARS-CoV-2 spike protein. The importance of studying the SARS-CoV-2 infection pathway is underscored by the widespread impact of the COVID-19 pandemic and the need to prevent future outbreaks. The spike protein initiates infection by opening its RBD, allowing it to attach to human angiotensin-converting enzyme 2 (ACE2) receptors and infiltrate the human cell. Using replica exchange umbrella sampling (REUS), we measured the free energy change accompanying this opening of the receptor binding domain (RBD). We thoroughly explored how glycans influence this transition using hydrogen bond and contact analysis beyond their passive role as protective shields against immune detection. Our findings provided insights into the exposure of epitopes to known antibodies, shedding light on potential therapeutic avenues against the virus and informing the design of future vaccines and drugs against other coronavirus [17].

References

1. J.M. Berg, J.L. Tymoczko, L. Stryer, G.J.J. Gatto, *Biochemistry* (International edition), 7th edn. (W.H. Freeman, New York, 2012), ISBN: 9781429276351
2. D.E. Koshland, Application of a theory of enzyme specificity to protein synthesis. Proc. Natl. Acad. Sci. U. S. A. **44**(2), 98–104 (1958)
3. J. Monod, J. Wyman, J.P. Changeux, On the nature of allosteric transitions: A plausible model. J. Mol. Biol. **12**(1), 88–118 (1965)
4. M. Karplus, J.A. McCammon, Molecular dynamics simulations of biomolecules. Nat. Struct. Biol. **9**(9), 646–652 (2002)
5. H. Frauenfelder, S.G. Sligar, P.G. Wolynes, The energy landscapes and motions of proteins. Science **254**(5038), 1598–1603 (1991)
6. M.J. Macielag, Chemical properties of antimicrobials and their uniqueness, in *Antibiotic Discovery and Development*, ed. by T.J. Dougherty, M. J. Pucci (Springer US, Boston, MA, 2012), pp. 793–820, ISBN: 978-1-4614-1400-1

7. M.R. Arkin, J.A. Wells, Small-molecule inhibitors of protein-protein interactions: progressing towards the dream. Nat. Rev. Drug Discov. **3**(3), 301–317 (2004)
8. C. Forrey, J.F. Douglas, M.K. Gilson, The fundamental role of flexibility on the strength of molecular binding. Soft Matter **8**(23), 6385–6392 (2012)
9. D. Shivakumar, Y. Deng, B. Roux, Computations of absolute solvation free energies of small molecules using explicit and implicit solvent model. J.Chem. Theory Comput. **5**(4), 919–930 (2009)
10. J.L. MacCallum, D.P. Tieleman, Interactions between small molecules and lipid bilayers. Curr. Top. Membr. **60**, 227–256 (2008)
11. C. Chipot, A. Pohorille, *Free Energy Calculations*, vol. 86 (Springer, 2007)
12. A. Shiber, T. Ravid, Chaperoning proteins for destruction: Diverse roles of Hsp70 chaperones and their co-chaperones in targeting misfolded proteins to the proteasome. Biomolecules **4**(3), 704–724 (2014)
13. S. Ojha, N. Patil, M. Jain, C. Kole, P. Kaushik, Probiotics for neurodegenerative diseases: a systemic review. Microorganisms, **11**(4), 1083 (2023)
14. K.M. Kuo, J. Liu, A. Pavlova, J.C. Gumbart, Drug binding to BamA targets its lateral gate. J. Phys. Chem. B **127**(34), 7509–7517 (2023)
15. B. Russell Lewis et al., Conformational restriction shapes the inhibition of a multidrug efflux adaptor protein. Nat. Commun. **14**(1), 3900 (2023)
16. D.L. Lynch, A. Pavlova, Z. Fan, J.C. Gumbart, Understanding virus structure and dynamics through molecular simulations. J. Chem. Theory Comput. **19**(11), 3025–3036 (2023)
17. Y.T. Pang, A. Acharya, D.L. Lynch, A. Pavlova, J.C. Gumbart, SARS-CoV-2 spike opening dynamics and energetics reveal the individual roles of glycans and their collective impact. Commun. Biol. **5**(1), 1170 (2022)
18. P.E. Wright, H.J. Dyson, Intrinsically disordered proteins in cellular signalling and regulation. Nat. Rev. Mol. Cell Biol. **16**(1), 18–29 (2015)
19. X. Su, Y.T. Pang, W. Li, J.C. Gumbart, J. Kelley, M. Torres, N-terminal intrinsic disorder is an ancestral feature of $G\gamma$ subunits that influences the balance between different $G\beta\gamma$ signaling axes in yeast. J. Biol. Chem. **299**(8), 104947 (2023)
20. F.A. Chao, R.A. Byrd, Protein dynamics revealed by NMR relaxation methods. Emerg. Top. Life Sci. **2**(1), 93–105 (2020)
21. D.E. Shaw et al., Anton 3: Twenty microseconds of molecular dynamics simulation before lunch, in *Int. Conf. High Perform. Comput. Netw.* (2021), pp. 1–11
22. NAMD performance. Accessed 19 Oct 2023, Theoretical and Computational Biophysics Group, University of Illinois at Urbana-Champaign
23. J. Schlitter, M. Engels, P. Krüger, Targeted molecular dynamics: a new approach for searching pathways of conformational transitions. J. Mol. Graph. **12**, 84–89 (1994)
24. Y. Sugita, Y. Okamoto, Replica-exchange molecular dynamics method for protein folding. Chem. Phys. Lett. **314**(1–2), 141–151 (1999)
25. J.J. Feld, D. Colledge, V. Sozzi, R. Edwards, M. Littlejohn, S.A. Locarnini, The phenylpropenamide derivative AT-130 blocks HBV replication at the level of viral RNA packaging. Antiviral Res. **76**(2), 168–177 (2007)
26. Y.T. Pang, A. Pavlova, E. Tajkhorshid, J.C. Gumbart, Parameterization of a drug molecule with a halogen σ-hole particle using fftk: Implementation, testing, and comparison. J.Chem. Phys. **153**(3), 164104 (2020)
27. I. Soteras Gutiérrez et al., Parametrization of halogen bonds in the CHARMM general force field: Improved treatment of ligand-protein interactions. Bioorg. Med. Chem. **24**(20), 4812–4825 (2016)
28. J.P. Renn, P.L. Clark, A conserved stable core structure in the passenger domain β-helix of autotransporter virulence proteins. Biopolymers **89**(5), 420–427 (2008)
29. Y.T. Pang, A.J. Hazel, J.C. Gumbart, Uncovering the folding mechanism of pertactin: A comparative study of isolated and vectorial folding. Biophys. J. **122**(14), 2988–2995 (2023)

Chapter 2
Parameterization of a Drug Molecule with a Halogen σ-Hole Particle Using ffTK: Implementation, Testing, and Comparison

Halogen atoms are widely used in drug molecules to improve their binding affinity for the receptor proteins. Many of the examples involve "halogen bonding" between the molecule and the binding site, which is a directional interaction between a halogen atom and a nucleophilic atom. Such an interaction is induced by an electron cloud shift of the halogen atom toward its covalently bonded neighbor to form the σ-bond, leaving a small electrostatic positive region opposite to the bond called the σ-hole." To mimic the effect of the σ-hole in the CHARMM non-polarizable force field, recently CGenFF added a positively charged massless particle to halogen atoms, positioned at the opposite side of the carbon–halogen bond. This particle is referred to as a lone pair (LP) particle because it uses the lone pair implementation in the CHARMM force field. Here, we have added support for LP particles to ffTK, an automated force field parameterization toolkit widely distributed as a plugin to the molecular visualization software VMD. We demonstrate the updated optimization process using an example halogenated drug molecule, AT130, which is a capsid assembly modulator targeting the hepatitis B virus. Our results indicate that parameterization with the LP particle significantly improves the accuracy of the electrostatic response of the molecule, especially around the halogen atom. Although the inclusion of the LP particle does not produce a prominent effect on the interactions between the molecule and its target protein, the protein–ligand binding performance is greatly improved by optimization of the parameters. Reprint from Pang, Y. T.; Pavlova, A.; Tajkhorshid, E.; Gumbart, J. C. "Parameterization of a drug molecule with a halogen σ-hole particle using ffTK: Implementation, testing, and comparison." *J Chem Phys*, vol. 153, no. 16, pp. 164104, 2020, with the permission of AIP Publishing.

Y. T. (Andrew) Pang, *Biomolecules' Conformational Changes Studied by Simulations and Enhanced Sampling*, Springer Theses,
https://doi.org/10.1007/978-3-031-70602-8_2

2.1 Introduction

Halogen atoms, such as chlorine and bromine, are frequently incorporated into drug molecules, enhancing binding affinity and selectivity towards their biomolecular targets [1–3]. Central to this enhancement is halogen bonding, a weak yet directionally specific interaction akin to hydrogen bonding, where a halogen atom engages electrophilic partners [4–7]. This bonding is facilitated by the formation of a σ-hole, a region of positive electrostatic potential created by the asymmetric distribution of the electron cloud along the axis of the covalent bond attaching the halogen to the molecule [1, 8]. The size of the σ-hole correlates with the halogen's polarizability and enables not only 'head-on' interactions with electrophilic atoms (halogen bond acceptors) but also 'side-on' interactions with electrophobic atoms (hydrogen bond donors), enhancing molecular interactivity (Fig. 2.1d). Moreover, the substantial van der Waals radii of common halogens contribute to additional hydrophobic interactions within the binding pocket [9, 10].

Historically, major molecular dynamics (MD) force fields like CHARMM [11], AMBER [12], OPLS [13, 14], and GROMOS [15] modeled electrostatic interactions using single point charges at the centers of atoms, neglecting the anisotropy of halogen substituents. This simplification limited their accuracy in reproducing the properties of halogenated molecules [16]. Responding to the growing interest in halogenated ligands, developers have updated these force fields by adding a positively charged virtual particle to the halogen atoms. This modification aims to emulate the effect of the σ-hole, enhancing the realism of simulations [17–21]. In the CHARMM General Force Field (CGenFF) model, for instance, a new σ-hole particle (LP) is rigidly attached opposite the covalent bond of its halogen host, ensuring consistent orientation (Fig. 2.1b) [17]. Although this model has been extensively validated with halogenated benzenes [17, 22, 23], its application to complex drug molecules, particularly those where halogens are not linked directly to phenyl rings, remains underexplored.

To enhance the parameterization of halogenated compounds not yet addressed by the recent CGenFF updates, we have modified the Force Field Toolkit (ffTK) [24] to integrate the LP particle into CGenFF 4.0, allowing for accurate representation of halogens. ffTK has been instrumental in simplifying the parameter optimization process across various systems over the years [25–31]. Here, we showcase the capabilities of our updated implementation through the parameterization of AT130, a brominated phenylpropenamide derivative targeting the hepatitis B virus (HBV) nucleocapsid. This molecule promotes the premature formation of empty viral capsids by accelerating nucleocapsid assembly [32]. The use of capsid-assembly modulators (CAMs) was shown to be a promising orthogonal approach to the presently approved therapies for eliminating the HBV infection [32–40]. AT130 was chosen due to its clinical significance and unique structure, characterized by a bromine atom linked to an sp2 carbon within an alkene bond, and juxtaposed with an amide and a methoxyphenyl group (Fig. 2.1a). To the best of our knowledge, this type of bonding has not been previously parameterized by CGenFF. In addition, this

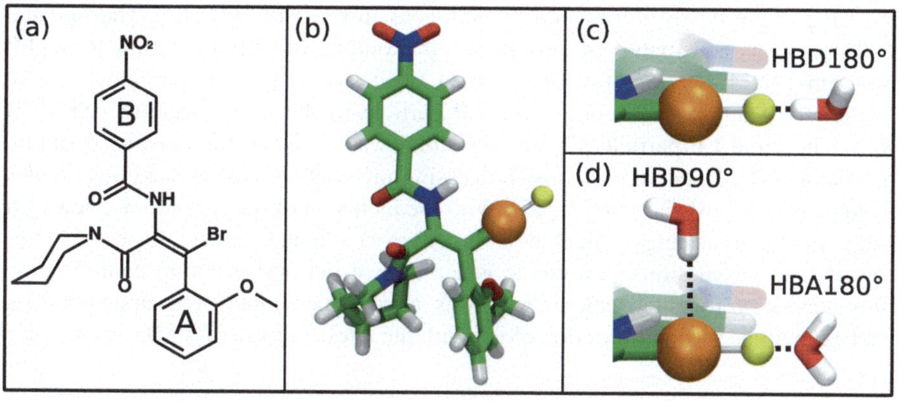

Fig. 2.1 AT130. (**a**) Structural formula. (**b**) 3D structure. In CGenFF 4.0, a positively charged massless particle (yellow sphere) is added to halogens (X), which in our case, is a bromine atom (orange sphere). This particle simulates the effect of the σ-hole at the opposite side of a C-X σ-bond. (**c**) Traditionally, halogens were treated solely as an electrophilic atom, and the water interaction energy was measured by a hydrogen-bond-donating water molecule placed parallel to the C-X bond (HBD180°). (**d**) To account for the anisotropy of halogens caused by the σ-hole, two new water orientations were added: one measuring the interaction energy by a hydrogen-bond-accepting water molecule parallel to the C-X bond (HBA180°), and one by a hydrogen-bond-donating water molecule perpendicular to the C-X bond (HBD90°). Reprinted from Pang, Y. T.; Pavlova, A.; Tajkhorshid, E.; Gumbart, J. C. "Parameterization of a drug molecule with a halogen σ-hole particle using ffTK: Implementation, testing, and comparison." *J Chem Phys*, vol. 153, no. 16, pp. 164104, 2020, with the permission of AIP Publishing

unusual bond is in a proximity to a methoxyphenyl group, making it an interesting candidate to test the new σ-hole implementation in CGenFF.

In this study, AT130 was parameterized using two approaches: one incorporating the LP particle and one without it. We evaluated both parameter sets alongside the original parameters sourced directly from the CGenFF program. Our comparison focused on three critical performance metrics: (1) the accurate portrayal of the electrostatic response across the molecule's surface, (2) the precise sampling of relevant molecular conformations, and (3) the molecular stability within the target protein's binding pocket. The refined parameters, validated through these metrics, are now publicly available at https://pubs.aip.org/jcp/article-supplement/200370/zip/164104_1_supplements/.

2.2 Methods

2.2.1 Implementation

ffTK [24] is a Tcl package distributed as a plugin of the program Visual Molecular Dynamics (VMD) [41]. The ffTK package was updated and expanded to support the LP particle newly introduced for halogens in CGenFF 4.0 [17]. The updated ffTK package is capable of processing the output file of the online CGenFF program [42, 43], computing the correct positions of the LP particles relative to their host atoms, and adding the LP particle to the corresponding PSF/PDB files. The same LP-particle-placing algorithm was added to the existing quantum mechanical (QM) calculation post-processing procedures used to read output files from Gaussian [44] and ORCA [45]. With the inclusion of the LP particle, new QM water interaction energy calculations at different orientations are suggested (See Results and Discussion). Automatic generation of the QM program configuration files necessary for such calculations was also implemented. The updated ffTK package will be available for download with the latest version of VMD 1.9.4.

2.2.2 Fitting of Force Field Parameters

The CHARMM [11] potential energy function consists of five bonded terms, including bond, angle, dihedral, improper and Urey-bradley potentials, and two non-bonded terms, including electrostatic and Lennard-Jones potentials (Eq. (2.1)).

$$V = \sum_{bonds} K_b(b-b_0)^2 + \sum_{angles} K_\theta(\theta-\theta_0)^2 + \sum_{dihedrals} K_\psi(1+\cos(n\psi-\delta))$$

$$+ \sum_{impropers} K_\phi(\phi-\phi_0)^2 + \sum_{Urey-Bradley} K_{UB}(r_{1,3}-r_{1,3;0})^2$$

$$+ \sum_{nonbonded} \frac{q_i q_j}{4\pi D r_{ij}} + \varepsilon_{ij}\left[\left(\frac{R_{min,ij}}{r_{ij}}\right)^{12} - 2\left(\frac{R_{min,ij}}{r_{ij}}\right)^{6})\right] \quad (2.1)$$

For this study, AT130 was uploaded to the CGenFF program [42, 43] (cgenff.umaryland.edu), and two sets of parameters were derived: one from the legacy version 1.0.0 without the LP particle (CGenFF parameters) and one from the updated version 2.2.0 incorporating the LP particle (CGenFF-LP parameters). CGenFF generates parameters by analogy and assesses them with a penalty score to indicate fit; a score over 10 indicates a poor match and further optimization is recommended [42].

Using ffTK, we optimized all charges, bonds, angles, and dihedral parameters with a penalty score exceeding 10. ffTK also facilitated the preparation of QM calculation configuration files, aiming to align molecular mechanical (MM) characteristics closely with the QM results. QM and MM calculations were conducted using Gaussian 16 [44] and NAMD 2.12-2.14 [46], respectively. All QM parameter optimizations were performed using the molecular geometry obtained by energy minimization at the MP2(6-31G*) level of theory. A lower-energy geometry was found during the first round of parameterization and a second round was initiated based on this new geometry. The data presented in this article is based on the latter round of parameterization. All parameter fittings were done first using simulated annealing and followed by iterations of optimization using the downhill simplex method [47, 48]. Downhill optimizations were repeated until convergence, meaning no parameter sets were found with better scores in two consecutive iterations.

Partial atomic charges were determined based on each atom's interaction with a TIP3P water molecule, ensuring compatibility with the CHARMM force field [11]. For each partial atomic charge to be fitted, a water molecule was placed at an ideal orientation for donating (accepting) hydrogen bond to (from) the atom of interest. Specifically, when acting as a hydrogen bond donor, the hydrogen of the water molecule was oriented towards the target atom, and as a hydrogen bond acceptor, the oxygen of the water faced the atom. The most energetically favorable interaction distance was established via qm energy minimization at the HF(6-31G*) level of theory. Nitrogens and oxygens were treated exclusively as hydrogen bond acceptors. Carbons, depending on their hybridization, engaged differently. Non-sp3 carbons were evaluated for direct water interactions in both donor and acceptor orientations, choosing the orientation with the lower interaction energy for subsequent charge optimization. For sp3 carbons, interactions were calculated based on the one of the covalently-bonded hydrogens. In AT130, both oxygen atoms O1 and O2 (see atom names in Fig. 2.2a) are ketones; Therefore, an additional water interaction calculation was done for each of these atoms with the hydrogen of the water molecule pointing at an angle of 120° from the C=O bond to interact with the lone pair electrons of the ketone oxygens. In fact, only the calculation at 120° was used for O1 because the water molecule was too close to the sugar ring when moving in line with the C=O bond. For the bromine atom BR1, three water interaction calculations were performed: two of which were the usual orientations where we placed the water molecule parallel to the C-Br bond, one as a hydrogen bond donor (HBD180°) and one as an acceptor (HBA180°), and one additional scan with the hydrogen of the water molecule pointing at BR1 along a direction perpendicular to the C-Br bond (HBD90°) (Fig. 2.1c,d) [22]. The configuration files for all of the above QM water interaction calculations were automatically generated by ffTK.

Due to spatial constraints posed by the proximity of the sugar ring and the methoxyphenyl group to the oxygen atom O1 in AT130, placing a water molecule for interaction calculations with atoms N2 and C3 was problematic. To circumvent this, a reduced version of AT130 was created by substituting the methoxyphenyl group with an aliphatic hydrogen (Fig. A.1a), a technique previously applied in macrolide studies [26]. The partial charges of C1, C2, C3, C4/C8, C9, C16, N1, N2, O1, O2, BR1 and H22 were first optimized on the reduced compound. The charges for C9 and C10 were subsequently optimized using the full structure of AT130. Using the results from these QM calculations, ffTK was employed to refine these atomic charges to align with the MM minimized water interaction energies and distances, and the dipole moment of the whole molecule with the QM calculations at the weight of energy:distance:dipole = 2:1:1. Each QM interaction energy calculation was equally weighted, except for those involving O2, and the HBD180° and HBA180° interactions with BR1, where weights were halved to ensure a balanced optimization across all atoms. Despite there are three distinct QM calculations for BR1, their weights were not reduced below 0.5 to prevent ffTK from underfitting these in comparison to other calculations. As the QM energy calculations were performed in the gas phase, adjustments were made to better represent aqueous conditions: interaction distances were decreased by 0.2 Å and energies were scaled up by a factor of 1.16 to account for polarization effects in the aqueous phase [11].

Bond and angle parameters were optimized based on the Hessian matrix calculated in internal coordinates at the MP2(6-31G*) level of theory [11]. However, due to the high computational demands of such frequency calculations, typically feasible only for compounds with fewer than 40 atoms. Given that AT130 comprises 53 atoms, we employed a reduced compound for these calculations, in which the nitrophenyl group and the sugar ring were substituted with a methyl group and a dimethyl amide group, respectively (Fig. A.1b). This adjustment reduced the number of atoms to 38, while preserving all the bonds and angles that we aimed to optimize. With the Hessian derived from this reduced compound, ffTK refined the bond and angle parameters to optimally match the MM geometry and the potential energy surface (PES) with the QM data. The weights for energy and geometry in the optimization process were set at a ratio of 2:1 to balance their contributions to the overall penalty score. During the optimization process, we noted occasional significant deviations in the equilibrium values of angle parameters, particularly those centered at atom N1, from the angles observed in the QM-minimized geometry. Although these parameters could replicate the QM geometry in MM minimization, they often resulted in unrealistic distortions during equilibrium simulations. Therefore, we implemented a corrective measure, resetting any angle parameters that drifted outside the range of 110° to 130° back to 120° to maintain structural integrity during simulations.

Lastly, the dihedral parameters were fitted by aligning the MM generated PES with that derived from relaxed dihedral QM scans at the MP2(6-31G*) level. Dihedrals were systematically rotated ±90° in 15° increments. Exceptions included the dihedrals C9-C10-C11-O03, C3-N2-C4-C5, and C3-N2-C8-C7 (See Fig. A.2 for atom names), which are part of ring structures that limit their rotation, hence they were rotated by only ±60° in 10° steps. The dihedral C1-N1-C2-C9 was rotated through a full 360° in 15° increments due to its flexibility and complex energy landscape later observed in MD simulations. The dihedral fitting process was executed in three stages. Initially, subsets of dihedral parameters were individually fitted, testing various combinations of multiplicities. Only those multiplicities essential for aligning the MM energy minima with the QM PES were retained to prevent overfitting [26]. Subsequently, all dihedrals were optimized collectively, with the phase shift for n = 2 multiplicities around the C2=C9 double bond fixed at 180°. The dihedral force constant was capped at 7 kcal/mol, an increase from the default 3 kcal/mol, to accommodate the potentially higher force constants required by the C2=C9 double bond. The energy cutoff began at 12 kcal/mol to encompass all energy minima on the PES, and was later reduced to 7 kcal/mol during the final optimization phases to minimize the root mean square errors of the most probable conformations. Finally, due to discrepancies in the MM sampling distribution of the flexible dihedrals around the C2-N1 bond, these parameters were reoptimized independently from the others. The energy cutoff for this adjustment started at 10 kcal/mol and was subsequently reduced to 5 kcal/mol, refining the alignment of these specific dihedrals with their expected energy profiles.

The QM energies of the three metastable states in an aqueous solvent were determined through a series of steps. Initially, three local energy minima identified from the dihedral scan of C1-N2-C2-C9 were isolated. The geometries of these minima were then optimized via QM energy minimization at the MP2(6-31G*) level of theory, establishing them as representative of the metastable states. Subsequently, the energy of each optimized structure was calculated using a single-point energy computation at the same MP2(6-31G*) level, incorporating a solvent reaction field simulated by the SMD model [49].

2.2.3 *Simulations*

All MD simulations were conducted using NAMD 2.12-2.14 [46], using the CHARMM36m protein force field [50] and the TIP3P water model [14]. For non-optimized AT130 parameters, CGenFF 3.0.1 and 4.3 (Ref. [11]) were used for simulations without and with the LP particle, respectively. The structure was constructed from the Protein Data Bank (PDB) structure 4G93. This is a structure of an assembled virus capsid of HBV with bound AT130, containing 240 copies of the capsid protein. Due to limitations of computational resources, we chose to simulate the asymmetric unit, a dimer of dimers, with AT130 bound at the interface between the two dimers. For the simulations of AT130 in water, the AT130 structure was extracted from the PDB and its geometry and parameters were optimized by ffTK [24] (See previous section). The molecule was then put into a cubic water box of side length 40 Å. The system was minimized for 20,000 steps and ran for 100 ns at 300 K and 1 atm. The simulation was performed four times, each with a different set of parameters for AT130: two using the original parameters obtained from the legacy and latest version of CGenFF program [42, 43], and two using the ffTK optimized parameter from each of them. For the protein simulation, the HBV capsid protein and AT130 were solvated in a box of dimensions $130 \times 150 \times 100$ Å. The system was minimized in two rounds for 2000 steps each. In the first round, we minimized the positions of water and ions while restraining the protein and the ligand. In the second round, the full system was minimized without any restraints. We previously showed that this two-step minimization can improve the stability of small compounds in their binding site during MD simulations [26]. The water molecules and ions were then equilibrated for 0.5 ns. After that, the system was run for 25 ns at 310 K and 1 atm with the protein backbone restrained. Such restraints were used to avoid potential conformational changes when the protein structure taken from an assembled virus capsid was simulated as a free tetramer. The simulation was performed five times for each of the four different set of parameters for AT130 as described.

The temperature and pressure of all simulations were maintained by Langevin dynamics and the Langevin piston method, respectively [51]. All simulations were performed under periodic boundary condition with a cutoff at 12 Å for short-range electrostatic and Lennard-Jones interactions, the latter with a force-based switching function beginning at 10 Å. Long-range electrostatic interactions were calculated using the particle-mesh Ewald (PME) method [52]. System setup, analysis and visualization were performed using VMD [41].

Root mean square deviation (RMSD) of AT130 was measured as the average distance of the heavy atoms of AT130 from the crystal structure. Solvent accessible surface area (SASA) was calculated using a solvent radius of 1.4 Å. The hydrophobic contact between AT130 and the protein was calculated by counting the number of carbon atoms from the protein that were within 4.5 Å of the hydrophobic region of AT130, which includes all carbon atoms on the two phenyl groups, the sugar ring and the methoxy group, the bromine atom and the carbon connected to it.

2.3 Results and Discussion

The initial force field parameters for AT130 were sourced from the CGenFF program [42, 43], with all parameters showing a penalty score greater than 10 subsequently optimized using ffTK (See Sect. 2.2 for the detailed workflow). Here, we focus on two prominent optimization steps that were most affected by the addition of the LP particle: (1) partial atomic charge optimization and (2) dihedral angle parameters optimization.

2.3.1 Charge Optimizations and Comparisons

Interaction energies were assessed by probing AT130 with a TIP3P water molecule from specified directions in MM calculations. These energies were then contrasted with those derived from QM calculations. Using ffTK, partial charges on selected atoms were meticulously adjusted to align the MM interaction energies and distances with those from QM calculations, as well as to accurately reflect the dipole moment of the entire molecule (Fig. 2.2).

Traditionally, halogen-water interaction energies were assessed with the water molecule serving as a hydrogen bond donor. This setup involved pointing one of the water's hydrogens toward the halogen (X), aligning its O-H bond with the C-X bond (HBD180°) (Fig. 2.1c). The introduction of the LP particle in CGenFF aimed to more accurately represent the anisotropic electrostatic responses of halogen atoms. This includes their interactions with hydrogen bond acceptors approached parallel to the C-X bond and hydrogen bond donors approached perpendicular to the bond.

With reference to previous works [17, 22], we incorporated two additional water probing orientations to better gauge the accuracy of MM calculations in capturing

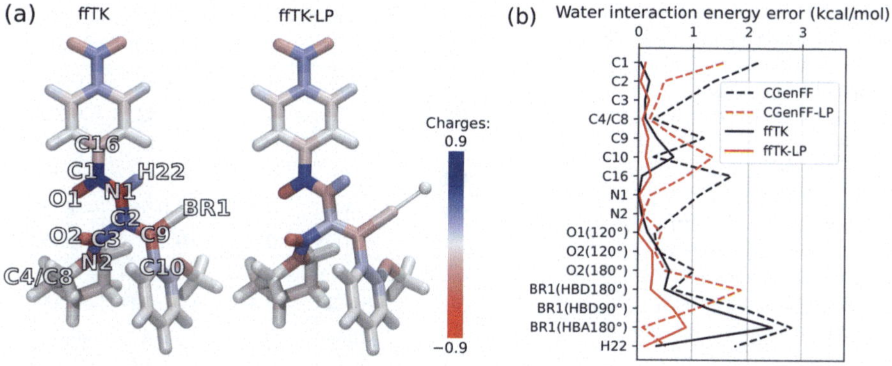

Fig. 2.2 Optimization of partial charges. (**a**) The optimized charges with and without LP are shown as a color scale from red (negative) to blue (positive) on the 3D structure. Names of the atoms with their partial atomic charge being optimized are shown. Without the LP particle (left), the charge of BR1 was set to almost 0 due to its attraction to both hydrogen bond donors and acceptors in QM calculations. C9 was then forced to be highly negative to compensate for the missing negative charge of BR1. The effect propagated to all other atoms being optimized. (**b**) MM water-interaction-energy error when compared to the QM data. While the ffTK parameters gave a much better agreement with the QM energies than CGenFF for most atoms, the water interactions with BR1 can only be improved by the introduction of the LP particle. Reprinted from Pang, Y. T.; Pavlova, A.; Tajkhorshid, E.; Gumbart, J. C. "Parameterization of a drug molecule with a halogen σ-hole particle using ffTK: Implementation, testing, and comparison." *J Chem Phys*, vol. 153, no. 16, pp. 164104, 2020, with the permission of AIP Publishing

these halogen and hydrogen bonding effects, and to fine-tune the partial charges for optimal alignment with QM results (Fig. 2.1d). In the first new orientation, the water molecule remained a hydrogen bond donor but approached the halogen from a direction perpendicular to the C-X bond (HBD90°). In the second orientation, the water acted as a hydrogen bond acceptor, with its oxygen directed toward the halogen along the C-X bond (HBA180°).

The partial charges derived from the latest version of the CGenFF program (CGenFF-LP), which incorporates the LP particle, closely resembled those from its legacy version (CGenFF). Notable exceptions included the charges of C3 and BR1, which increased by 0.121 and decreased by 0.054, respectively; all other differences remained below 0.05 (Table 2.1). However, following optimization with ffTK, significant disparities emerged between the charge sets with and without the LP particle (ffTK-LP and ffTK, respectively). In the absence of the LP particle's small positive charge, BR1 in the ffTK parameter set was adjusted to nearly neutral, reflecting its reduced attraction to the hydrogen-bond-accepting water molecule. As a consequence, the neighboring C9 compensated for the absence of BR1's negative charge, resulting in an excessively negative fitting. This adjustment affected the charge balance of other atoms involved in the optimization process (Fig. 2.2a).

The introduction of the LP particle significantly reduced the interaction-energy errors, as evidenced by a comparison of water interaction energies across different parameter sets (Fig. 2.2b). Without the LP particle, the interaction-energy errors for

Table 2.1 Partial charges obtained from CGenFF and after optimization by ffTK with and without the LP particle. Note that the charge on LP was not included in the optimization. Reprinted from Pang, Y. T.; Pavlova, A.; Tajkhorshid, E.; Gumbart, J. C. "Parameterization of a drug molecule with a halogen σ-hole particle using ffTK: Implementation, testing, and comparison." *J Chem Phys*, vol. 153, no. 16, pp. 164104, 2020, with the permission of AIP Publishing

	CGenFF		ffTK	
Atom	No LP	With LP	No LP	With LP
C1	0.412	0.453	0.915	0.662
C2	0.105	0.102	0.582	0.224
C3	0.384	0.505	0.540	0.531
C4/C8	0.139	0.095	0.062	0.090
C9	0.239	0.253	−0.551	−0.325
C10	−0.005	−0.001	0.251	0.255
C16	−0.068	−0.109	−0.264	−0.218
N1	−0.549	−0.559	−0.721	−0.358
N2	−0.499	−0.513	−0.479	−0.474
O1	−0.440	−0.441	−0.662	−0.589
O2	−0.514	−0.513	−0.554	−0.522
BR1	−0.140	−0.194	−0.020	−0.200
H22	0.325	0.325	0.367	0.329
LP	-	0.050	-	0.050

BR1 with the CGenFF parameters were substantial, at 1.9 kcal/mol for HBD90° and 2.8 kcal/mol for HBA180°. These errors indicated a significant discrepancy in replicating the interactions of bromine with both hydrogen bond donors and acceptors. Despite optimization with ffTK, the errors for HBD90° and HBA180° only marginally improved to 1.3 and 2.4 kcal/mol, respectively.

However, with the inclusion of the LP particle in CGenFF-LP, the interaction-energy errors dramatically decreased to 1.1 and 0.1 kcal/mol, respectively. Further optimization with ffTK (ffTK-LP) adjusted these errors to 0.6 and 0.9 kcal/mol, respectively, enhancing the accuracy for HBD90° while slightly sacrificing the accuracy for HBA180°. This adjustment enabled the ffTK-LP charges to more effectively balance BR1's dual role as both a halogen bond donor and a hydrogen bond acceptor. Notably, the ffTK-LP was the only parameter set in which all water-interaction-energy errors were reduced to less than 1 kcal/mol, underscoring its improved performance in simulating realistic halogen interactions.

2.3.2 *Dihedral Optimizations*

Variations in the distribution of partial charges not only influence the electrostatic response of the molecule to its environment but also impact intramolecular electrostatic interactions. In the case of AT130, a highly flexible dihedral around the C2-N1 bond facilitates the existence of three distinct metastable states: (1) H22 engages in favorable electrostatic interactions with BR1 and C9, (2) H22 rotates clockwise to establish a favorable electrostatic interaction with O2, and (3) H22 turns anticlockwise, distancing itself from both BR1 and O2 (Fig. 2.3b).

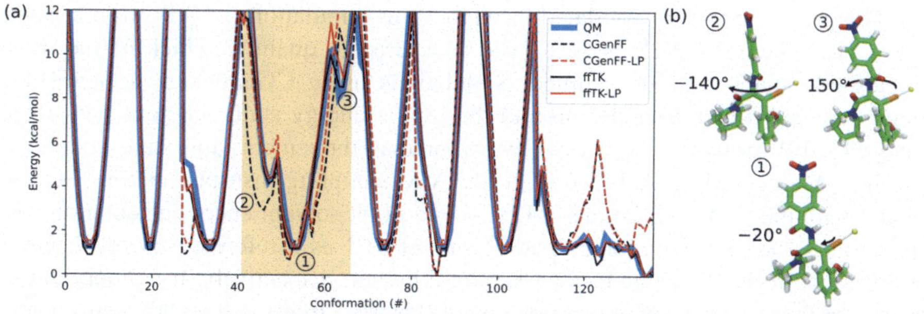

Fig. 2.3 Optimization of dihedrals. (**a**) PES of AT130 when rotating the fitted dihedrals. Both the ffTK and ffTK-LP parameters drastically improved the fit to the QM energy surface when compared to the CGenFF and CGenFF-LP parameters. The most challenging dihedrals to fit were the ones around the C2-N1 bond (shaded region). (**b**) Three metastable states were found when rotating around this bond, in which H22 points (1) in the direction of BR1, (2) in the direction of O2, or (3) away from both of them. Reprinted from Pang, Y. T.; Pavlova, A.; Tajkhorshid, E.; Gumbart, J. C. "Parameterization of a drug molecule with a halogen σ-hole particle using ffTK: Implementation, testing, and comparison." *J Chem Phys*, vol. 153, no. 16, pp. 164104, 2020, with the permission of AIP Publishing

This flexibility introduces significant complexity to the optimization of dihedral parameters, influencing the overall molecular behavior and interaction dynamics.

Given the complex energy landscape associated with the flexible dihedral, precise tuning of the dihedral parameters is critical. Using ffTK, dihedral parameters from both CGenFF and CGenFF-LP were optimized to best reproduce the potential energy surface (PES) derived from QM calculations when rotating the relevant dihedral angles in MM simulations. Due to the intricate energy profile around the C-N bond, dihedrals around the central atom C2 were initially addressed to identify the minimal set of multiplicities necessary to accurately reproduce the QM PES. As previously demonstrated, minimizing the number of multiplicities in dihedral terms is essential to prevent overfitting [26].

After comprehensive testing, an additional multiplicity was integrated into the dihedral parameters for CG2DC1 CG2D1O NG2S1 CG2O1 (C9-C2-N1-C1), altering the multiplicity set from $n = 1, 2$ to $n = 1, 2, 3$. This modification shifted the MM energy minimum to align with the QM scans (Fig. A.3). Subsequently, all dihedrals with high penalty scores were optimized together. The fitting process began with an energy cutoff set to 12 kcal/mol to encompass all three metastable states observed in the QM scans. Once the downhill optimization converged, the energy cutoff was reduced to 7 kcal/mol to refine the fit for the low energy states that are more frequently encountered in equilibrium simulations. Ultimately, the final parameters of both ffTK and ffTK-LP achieved commendable consistency with the QM PES (Fig. 2.3a). The root mean square errors were calculated to be 0.381 and 0.349, respectively, with the energy cutoff of 7 kcal/mol, illustrating effective parameter tuning.

The parameters were validated through 10 ns simulations of AT130 in a water box. The dihedral C9-C2-N1-H22 was measured to quantify which of the three metastable states AT130 occupied. Simulations using CGenFF and CGenFF-LP only sampled dihedral angles around the lowest energy state, whereas both ffTK and ffTK-LP simulations successfully explored all three metastable states identified in the QM scans (Fig. A.4). To align the MM sampling distribution computed in water with the QM calculations, a QM single-point solvent energy calculation was performed for each metastable state. Using state 1 as a reference, states 2 and 3 were found to be +2.19 and +4.63 kcal/mol higher, respectively. It is important to note that these single-point energies may differ from those in the QM scans, as the former were conducted with a solvent potential while the latter are conducted in a vacuum. By applying the Boltzmann factor $p_i/p_j = \exp(\Delta E/kT)$, we estimate the probability of visiting state 1, 2 and 3 to be 97.49%, 2.47% and 0.04%, respectively. Despite closely matching the QM PES, the ffTK and ffTK-LP simulations did not accurately reproduce the expected population distribution of the three metastable states. Especially in ffTK-LP simulations, state 2 was sampled as frequently as, or even slightly more than, state 1. This anomaly could be attributed to the partial charge of C9 in ffTK-LP being 0.226*e* less negative than in ffTK, which reduced the electrostatic attraction between C9 and H22, thus making state 1 less favorable. Additionally, both ffTK and ffTK-LP simulations tended to oversample state 3.

To address discrepancies between MM simulations and QM predicted values, a focused refinement was implemented on the dihedral parameters around the C2-N1 bond, treated separately from other dihedrals. According to CGenFF guidelines used by ffTK, QM-optimized geometries from dihedral scans should not be directly employed for MM dihedral term fitting. This is due to potential misalignments between QM and MM calculated values for bond and angle terms despite previous optimizations, which can lead to exaggerated MM energy calculations and potentially distort the energy contributions from the dihedral terms. To mitigate this issue, ffTK starts by using structures obtained from QM scans and performs a MM energy minimization on the structure with the dihedrals currently being optimized are intentionally kept fixed. This practice prevents the use of potentially unreliable dihedral parameters that could result in nonphysical geometries and inaccurately calculated energies. Keeping these dihedrals fixed ensures that the minimized MM structures do not deviate drastically due to unstable parameter values, maintaining the integrity of the comparison between MM and QM dihedral potentials. However, locking these dihedrals during minimization can also lead to structures that deviate from those observed in unrestrained equilibrium simulations. To circumvent this potential issue, after completing a full parameter set optimization, a small subset of dihedral parameters is refitted. This strategy allows us to avoid the danger of having distorted structures caused by unreliable dihedral parameters, yet take advantage of using a structure with fewer the dihedrals restrained, thus obtaining a more realistic MM PES.

The PES before and after refitting are illustrated in Fig. 2.4a. Initially, the MM PES generated with only the dihedrals around the C2-N1 bond locked indicated that our original ffTK and ffTK-LP parameters favored state 2. These dihedral

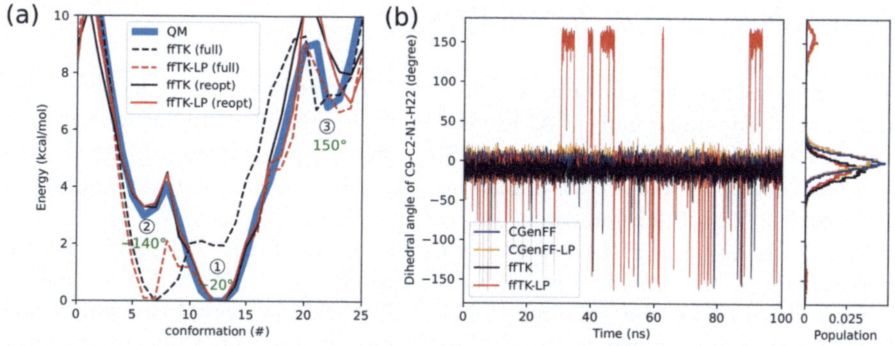

Fig. 2.4 Reoptimizing dihedrals about the C2-N1 bond. (**a**) PES of AT130 when rotating the dihedral C1-N1-C2-C9 after full parameter set optimization (dotted) and after further reoptimizing the dihedral parameters around the C2-N1 bond (solid). After the reoptimization, the energy minima of ffTK and ffTK-LP were shifted from state 2 back to state 1, but sacrificed some accuracy for state 3. The green labels are the C9-C2-N1-H22 dihedral angles measured at the corresponding states. (**b**) Trajectory and populations for the dihedral angle of C9-C2-N1-H22 in a 100-ns simulation in water using different parameter sets. While both CGenFF and CGenFF-LP parameters only sampled around 0°, ffTK parameters were able to sample the two lower energy states at −20° and −140°. ffTK-LP parameters even sample state 3 at 150°, although it was oversampled. Reprinted from Pang, Y. T.; Pavlova, A.; Tajkhorshid, E.; Gumbart, J. C. "Parameterization of a drug molecule with a halogen σ-hole particle using ffTK: Implementation, testing, and comparison." *J Chem Phys*, vol. 153, no. 16, pp. 164104, 2020, with the permission of AIP Publishing

parameters underwent initial refitting with an energy cutoff set at 10 kcal/mol to encompass all three minima of the PES. Subsequently, the cutoff was reduced to 5 kcal/mol to achieve a more precise fit for the states visited more frequently. The refitted ffTK and ffTK-LP parameters yielded root mean square errors of 0.356 and 0.294, respectively, when measured with the final energy cutoff. Following the parameter adjustments, AT130 was subjected to 100-ns simulations in water for each parameter set. Both ffTK and ffTK-LP simulations produced a population distribution between states 1 and 2 that was reasonably accurate. Specifically, the percentage of frames in state 2 was calculated to be 0.7% for ffTK and 3.4% for ffTK-LP, closely aligning with predictions based on the Boltzmann factor derived from the QM energy differences between these states in solvent. Additionally, the ffTK-LP simulation also visited state 3, albeit at a significantly higher frequency of 14.29%, compared to the predicted population. Conversely, state 3 was never sampled in the 100-ns ffTK simulation, which is consistent with its low predicted probability. Its agreement with the QM data could be confirmed with enhanced sampling methods, such as umbrella sampling, which is outside the scope of our current study.

2.3.3 Simulations in Protein Pocket

To further assess the performance of each parameter set, we conducted five individual 25-ns simulations of AT130 within its target protein binding pocket, with the protein backbone restrained. The RMSD from the crystal structure of AT130 was measured and compared across simulations (Fig. 2.5). The average RMSD of AT130 from the last 5 ns of the simulations was calculated to be 4.0 $\pm$ 1.0 Å, 3.6 $\pm$ 1.7 Å, 1.6 $\pm$ 0.3 Å, and 1.7 $\pm$ 0.4 Å for the CGenFF, CGenFF-LP, ffTK, and ffTK-LP parameter sets, respectively. These results clearly demonstrate that the parameters optimized with ffTK showed improved alignment with the crystal structure. Upon closer examination of the individual trajectories, it was noted that none of the CGenFF simulations maintained an RMSD below 2 Å during the last 5 ns. However, two of the CGenFF-LP simulations did achieve this benchmark, suggesting that the inclusion of the LP particle enhances the stability of AT130 within the binding pocket when using CGenFF parameters. Comparatively, the introduction of the LP particle had a marginal effect on RMSD outcomes between the ffTK and ffTK-LP simulations. All five ffTK simulations and four of the five ffTK-LP simulations concluded with an RMSD under 2 Å. Yet, in two of the ffTK-LP simulations (Sim3 and Sim4), AT130 consistently exhibited an RMSD greater than 2 Å for a significant portion of the simulation, differing from the other ffTK and ffTK-LP simulations.

Further analysis of the ffTK-LP simulations revealed recurring issues with oversampling the highest energy metastable state, state 3, similar to those observed in the in-water simulations. Notably, in two of the five protein pocket simulations (Sim1 and Sim4), AT130 was found in state 3, characterized by the C1-N1-C2-C9 dihedral being rotated to around 150°. Unlike in the in-water simulations, once adopted at the start of the simulations, the molecule maintained either state 1 or state 3 throughout the remaining trajectory. Intriguingly, the adopted conformation did not directly correspond to the RMSD values of AT130 from the crystal structure. For instance, Sim1, which consistently exhibited state 3, recorded the lowest average RMSD among the ffTK-LP simulations at 1.40 Å, while Sim4, also in state 3, had the highest RMSD at 2.33 Å (Table A.1).

In an effort to understand the differences between the CGenFF/CGenFF-LP and ffTK/ffTK-LP simulations, frames exhibiting high RMSD were visually inspected. It was observed that in many of the CGenFF and CGenFF-LP simulations, AT130 tended to move toward the exit of the pocket, whereas in the ffTK and ffTK-LP simulations, the ligand consistently remained near its crystal-structure position (Fig. 2.5b). This behavior was substantiated by measuring the SASA of AT130 in each set of simulations. Notably, AT130 maintained a low SASA, ranging from 50 to 100 Å^2, in all ffTK and ffTK-LP simulations. In contrast, the SASA in many of the CGenFF and CGenFF-LP simulations increased significantly, reaching up to 170 Å^2 (Fig. A.5). Furthermore, the number of hydrophobic contacts between AT130 and the protein was reduced from $\sim$65 to $\sim$50 in the CGenFF and CGenFF-LP simulations, while remaining consistent at $\sim$65 in all ffTK and ffTK-LP simulations (Fig. A.6). These findings clearly demonstrate that AT130

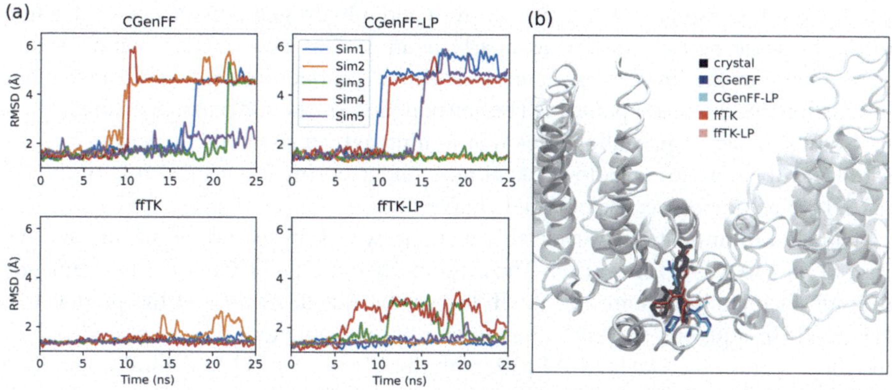

Fig. 2.5 Bound-state of AT130. (**a**) The RMSD of AT130 from its crystal-structure position when simulating AT130 in the binding pocket with the protein backbone restrained. While the RMSDs of AT130 are all well below 3.5 Å when using the ffTK or ffTK-LP parameters, many of the RMSD in simulations using CGenFF and CGenFF-LP parameters went over 4 Å. The inclusion of the LP particle did not significantly affect the average RMSD of AT130 in the binding pocket, showing a slight improvement in the CGenFF simulations, and a slight deterioration in the ffTK simulations. (**b**) Representative frames with an RMSD in the upper quarter of the RMSD distribution of each parameter set were overlad on top of each other. AT130 remained near its the crystal structure position when using the ffTK and ffTK-LP parameters but drifted towards the pocket exit for the CGenFF and CGenFF-LP ones. Reprinted from Pang, Y. T.; Pavlova, A.; Tajkhorshid, E.; Gumbart, J. C. "Parameterization of a drug molecule with a halogen σ-hole particle using ffTK: Implementation, testing, and comparison." *J Chem Phys*, vol. 153, no. 16, pp. 164104, 2020, with the permission of AIP Publishing

exhibits greater stability within the binding pocket when simulated with ffTK or ffTK-LP parameters compared to CGenFF or CGenFF-LP parameters. Interestingly, the introduction of the LP particle did not significantly affect the protein-binding affinity of AT130, suggesting that while the LP particle enhances certain aspects of the simulation, it does not notably influence binding stability.

2.4 Conclusions

We integrated support for the new LP particles in CGenFF 4.0 into the ffTK package, and showcased this enhancement through the optimization of force-field parameters for the HBV capsid-assembly modulator, AT130. The effectiveness of including the LP particle was evaluated by optimizing and comparing parameters with and without the LP particle using ffTK. The comparison extended to pre- and post-optimization parameters. Our findings clearly demonstrated that incorporating the LP particle significantly enhanced the congruence between the QM- and MM-calculated water interaction energies, particularly for the bromine atom. Without the LP particle, the MM parameters lacked a mechanism to accurately represent the attractive forces between the bromine atom and a hydrogen bond acceptor, leading to the bromine being modeled as nearly neutral. This necessitated compensatory adjustments in the charges of neighboring atoms, affecting the overall molecular interaction energies and polarity. The introduction of the LP particle enabled ffTK to establish a set of partial charges that minimized water-interaction-energy errors for all involved atoms to below 1 kcal/mol, underscoring the critical role of the LP particle in achieving accurate partial charge fittings.

Dihedral optimization is invariably a complex task in force-field fitting, and the case of AT130 was no exception. The original CGenFF and CGenFF-LP parameters did not adequately capture the flexibility of the dihedrals around the pivotal C2-N1 bond. In aqueous solvent simulations, these parameters only captured one of the three metastable states of AT130, with the C9-C2-N1-H22 dihedral population deviating slightly from the energy minimum value. Through a meticulous multi-step dihedral fitting process, we managed to achieve a satisfactory sampling distribution of the two lowest energy states using the ffTK and ffTK-LP parameters. Despite a good alignment between the QM and MM PES, both the ffTK and ffTK-LP parameters encountered challenges in reproducing the precise sampling distribution of the C2-N1 dihedrals. The ffTK parameters failed to sample, and the ffTK-LP parameters oversampled the highest energy metastable state. However, both ffTK parameter sets outperformed the CGenFF configurations in accurately sampling the two lower energy states, more closely mirroring the distributions predicted by QM. Additionally, our findings underscore the benefit of refitting a small subset of dihedral parameters post-comprehensive parameter fitting, significantly enhancing the efficacy of the force-field parameters.

Finally, we tested all four parameter sets through simulations in the target protein binding pocket. While the results were inconclusive regarding whether the LP particle enhances the protein-AT130 interactions, it is evident that parameter optimization significantly bolsters AT130's stability within the binding pocket. This observation is supported by metrics such as the RMSD of AT130 relative to the crystal structure, the SASA of AT130, and the number of hydrophobic contacts between AT130 and the protein. Visual analysis of the trajectories indicated that AT130 was more prone to lose contact with the protein and migrate towards the exit of the binding pocket when simulated with the CGenFF or CGenFF-LP parameters compared to the ffTK or ffTK-LP parameters. Consequently, we recommend conducting parameter optimization before running simulations on novel molecules that exhibit high penalty scores in the CGenFF program. In the case of AT130, there are no halogen bonds between its bromine atom and the target protein observed in the crystal structure, only non-specific hydrophobic interactions. Although further testing is required to generalize these findings, our results suggest that the addition of the LP particle neither enhances nor impacts the hydrophobic interactions of halogen substituents. Therefore, the efficacy of including the LP particle in improving the accuracy of binding affinity in simulations depends on the specific system being studied.

References

1. G. Cavallo et al., The halogen bond. Chem. Rev. **116**(4), 2478–2601 (2016)
2. S. Buchini, A. Buschiazzo, S.G. Withers, A new generation of specific Trypanosoma cruzi trans-sialidase inhibitors. Angew. Chem. Int. Ed. Engl. **47**(14), 2700–2703 (2008)
3. A.C. Leite et al., Synthesis, Cruzain docking, and in vitro studies of aryl-4-oxothiazolylhydrazones against Trypanosoma cruzi. ChemMedChem **2**(9), 1339–1345 (2007)
4. B.L. Grasberger et al., Discovery and cocrystal structure of benzodiazepinedione HDM2 antagonists that activate p53 in cells. J. Med. Chem. **48**(4), 909–912 (2005)
5. D.J. Parks et al., 1,4-Benzodiazepine-2,5-diones as small molecule antagonists of the HDM2-p53 interaction: discovery and SAR. Bioorg. Med. Chem. Lett. **15**(3), 765–770 (2005)
6. O. Fedorov et al., Specific CLK inhibitors from a novel chemotype for regulation of alternative splicing. Chem. Biol. **18**(1), 67–76 (2011)
7. L.A. Hardegger et al., Halogen bonding at the active sites of human cathepsin L and MEK1 kinase: efficient interactions in different environments. ChemMedChem **6**(11), 2048–2054 (2011)
8. Á.M. Montaña, The σ and π holes. the halogen and tetrel bondings: Their nature, importance and chemical, biological and medicinal implications. ChemistrySelect **2**(28), 9094–9112 (2017)
9. M.R. Scholfield, C.M. Zanden, M. Carter, P.S. Ho, Halogen bonding (X-bonding): a biological perspective. Protein Sci. **22**(2), 139–152 (2013)
10. K. Naumann, Influence of chlorine substituents on biological activity of chemicals. J. Prakt. Chem. **341**(5), 417–435 (1999)
11. K. Vanommeslaeghe et al., CHARMM general force field: A force field for drug-like molecules compatible with the CHARMM all-atom additive biological force fields. J. Comput. Chem. **31**(4), 671–690 (2010)
12. W.D. Cornell et al., A second generation force field for the simulation of proteins, nucleic acids, and organic molecules. J. Am. Chem. Soc. **117**(19), 5179–5197 (1995)
13. W.L. Jorgensen, J. Tirado-Rives, The OPLS [optimized potentials for liquid simulations] potential functions for proteins, energy minimizations for crystals of cyclic peptides and crambin. J. Am. Chem. Soc. **110**(6), 1657–1666 (1988)
14. W.L. Jorgensen, J. Chandrasekhar, J.D. Madura, R.W. Impey, M.L. Klein, Comparison of simple potential functions for simulating liquid water. J. Chem. Phys. **79**(2), 926–935 (1983)
15. W.R. Scott et al., The GROMOS software for biomolecular simulation: GROMOS05. J. Comput. Chem. **26**(16), 1719–1751 (2005)
16. A.N. Adluri, J.N. Murphy, T. Tozer, C.N. Rowley, Polarizable force field with a σ-hole for liquid and aqueous bromomethane. J. Phys. Chem. B **119**(42), 13422–13432 (2015)
17. I. Soteras Gutiérrez et al., Parametrization of halogen bonds in the CHARMM general force field: Improved treatment of ligand-protein interactions. Bioorg. Med. Chem. **24**(20), 4812–4825 (2016)
18. E. Harder et al., OPLS3: a force field providing broad coverage of drug-like small molecules and proteins. J. Chem. Theory Comput. **12**(1), 281–296 (2016)
19. M.A. Ibrahim, Molecular mechanical study of halogen bonding in drug discovery. J. Comput. Chem. **32**(12), 2564–2574 (2011)
20. M.A. Ibrahim, Molecular mechanical perspective on halogen bonding. J. Mol. Model. **18**(10), 4625–4638 (2012)
21. X. Mu et al., Modeling organochlorine compounds and the σ-hole effect using a polarizable multipole force field. J. Phys. Chem. B **118**(24), 6456–6465 (2014)
22. F.Y. Lin, A.D. MacKerell Jr., Do halogen–hydrogen bond donor interactions dominate the favorable contribution of halogens to ligand–protein binding? J. Phys. Chem. B **121**, 6813–6821 (2017)

23. F.Y. Lin, A.D. MacKerell Jr., Improved modeling of halogenated ligand-protein interactions using the drude polarizable and CHARMM additive empirical force fields. J. Chem. Inf. Model. **59**, 215–228 (2019)
24. C.G. Mayne, J. Saam, K. Schulten, E. Tajkhorshid, J.C. Gumbart, Rapid parameterization of small molecules using the Force Field Toolkit. J. Comput. Chem. **34**(32), 2757–2770 (2013)
25. J.C. Gumbart, M. Beeby, G.J. Jensen, B. Roux, Escherichia coli peptidoglycan structure and mechanics as predicted by atomic-scale simulations. PLoS Comput. Biol. **10**(2), e1003475 (2014)
26. A. Pavlova, J.C. Gumbart, Parametrization of macrolide antibiotics using the force field toolkit. J. Comput. Chem. **36**(27), 2052–2063 (2015)
27. L. Han et al., Transparent, adhesive, and conductive hydrogel for soft bioelectronics based on light-transmitting polydopamine-doped polypyrrole nanofibrils. Chem. Mater. **30**(16), 5561–5572 (2018)
28. Y.C. Chiang, O. Levsh, C.K. Lam, J.K. Weng, Y. Wang, Structural and dynamic basis of substrate permissiveness in hydroxycinnamoyltransferase (HCT). PLoS Comput. Biol. **14**(0), e1006511 (2018)
29. Q. Chen et al., Structural basis of neurosteroid anesthetic action on GABAA receptors. Nat. Commun. **9**(1), 3972 (2018)
30. J.A. Coleman et al., Serotonin transporter-ibogaine complexes illuminate mechanisms of inhibition and transport. Nature **569**(7754), 141–145 (2019)
31. Y.-C. Chiang, M.T. Wong, J.W. Essex, Molecular dynamics simulations of antibiotic ceftaroline at the allosteric site of penicillin-binding protein 2a (PBP2a). Isr. J. Chem. **60**, 754 (2020)
32. J.J. Feld, D. Colledge, V. Sozzi, R. Edwards, M. Littlejohn, S.A. Locarnini, The phenylpropenamide derivative AT-130 blocks HBV replication at the level of viral RNA packaging. Antiviral Res. **76**(2), 168–177 (2007)
33. K. Deres et al., Inhibition of hepatitis B virus replication by drug-induced depletion of nucleocapsids. Science **299**(5608), 893–896 (2003)
34. C. Bourne, et al., Small-molecule effectors of hepatitis B virus capsid assembly give insight into virus life cycle. J. Virol. **82**(20), 10262–10270 (2008)
35. W.E. Delaney et al., Phenylpropenamide derivatives AT-61 and AT-130 inhibit replication of wild-type and lamivudine-resistant strains of hepatitis B virus in vitro. Antimicrob. Agents Chemother. **46**(9), 3057–3060 (2002)
36. M.R. Campagna, et al., Sulfamoylbenzamide derivatives inhibit the assembly of hepatitis B virus nucleocapsids. J. Virol. **87**(12), 6931–6942 (2013)
37. Z. Zhou et al., Heteroaryldihydropyrimidine (HAP) and sulfamoylbenzamide (SBA) inhibit Hepatitis B virus replication by different molecular mechanisms. Sci. Rep. **7**, 42374 (2017)
38. O. Sari et al., Synthesis of sulfamoylbenzamide derivatives as HBV capsid assembly effector. Eur. J. Med. Chem. **138**, 407–421 (2017)
39. A.G. Cole, Modulators of HBV capsid assembly as an approach to treating hepatitis B virus infection. Curr. Opin. Pharmacol. **30**, 131–137 (2016)
40. L. Yang, F. Liu, X. Tong, D. Hoffmann, J. Zuo, M. Lu, Treatment of chronic hepatitis B virus infection using small molecule modulators of nucleocapsid assembly: Recent advances and perspectives. ACS Infect. Dis. **5**(5), 713–724 (2019)
41. W. Humphrey, A. Dalke, K. Schulten, VMD: visual molecular dynamics. J. Mol. Graph. **14**(1), 33–38 (1996)
42. K. Vanommeslaeghe, A.D. MacKerell Jr., Automation of the CHARMM General Force Field (CGenFF) I: bond perception and atom typing. J. Chem. Inf. Model. **52**(12), 3144–3154 (2012)
43. K. Vanommeslaeghe, E.P. Raman, A.D. MacKerell Jr., Automation of the CHARMM General Force Field (CGenFF) II: assignment of bonded parameters and partial atomic charges. J. Chem. Inf. Model. **52**(12), 3155–3168 (2012)
44. M. Frisch et al., Gaussian 16, revision A. 03 (Gaussian Inc., Wallingford CT, 2016)
45. F. Neese, The ORCA program system. Wiley Interdiscip. Rev. Comput. Mol. Sci. **2**(1), 73–78 (2012)

46. J.C. Phillips et al., Scalable molecular dynamics with NAMD. J. Comput. Chem. **26**(16), 1781–1802 (2005)
47. J.A. Nelder, R. Mead, A simplex method for function minimization. Comput. J. **7**(4), 308–313 (1965)
48. W.H. Press, S.A. Teukolsky, W.T. Vetterling, B.P. Flannery, *Numerical recipes in C*, 2nd edn. (Cambridge University Press, New York, 1992)
49. A.V. Marenich, C.J. Cramer, D.G. Truhlar, Universal solvation model based on solute electron density and on a continuum model of the solvent defined by the bulk dielectric constant and atomic surface tensions. J. Phys. Chem. B **113**, 6378–6396 (2009)
50. J. Huang et al., CHARMM36m: an improved force field for folded and intrinsically disordered proteins. Nat. Methods **14**(1), 71–73 (2017)
51. S.E. Feller, Y. Zhang, R.W. Pastor, B.R. Brooks, Constant pressure molecular dynamics simulation: the Langevin piston method. J. Chem. Phys. **103**(11), 4613–4621 (1995)
52. T. Darden, D. York, L. Pedersen, Particle mesh Ewald: An $N \cdot \log(N)$ method for Ewald sums in large systems. J. Chem. Phys. **98**(12), 10089–10092 (1993)

Chapter 3
Uncovering the Folding Mechanism of Pertactin: A Comparative Study of Isolated and Vectorial Folding

Autotransporters are a large family of virulence factors found in Gram-negative bacteria that play important roles in their pathogenesis. The passenger domain of autotransporters is almost always composed of a large β-helix, with only a small portion of it being relevant to its virulence function. This has led to the hypothesis that the folding of the β-helical structure aids the secretion of the passenger domain across the Gram-negative outer membrane. In this study, we used molecular dynamics simulations and enhanced sampling methods to investigate the stability and folding of the passenger domain of pertactin, an autotransporter from Bordetella pertussis. Specifically, we employed steered molecular dynamics to simulate the unfolding of the entire passenger domain as well as self-learning adaptive umbrella sampling to compare the energetics of folding rungs of the β-helix independently ("isolated folding") versus folding rungs on top of a previously folded rung ("vectorial folding"). Our results showed that vectorial folding is highly favorable compared with isolated folding; moreover, our simulations showed that the C-terminal rung of the β-helix is the most resistant to unfolding, in agreement with previous studies that found the C-terminal half of the passenger domain to be more stable than the N-terminal one. Overall, this study provides new insights into the folding process of an autotransporter passenger domain and its potential role in secretion across the outer membrane. Reprinted from Biophys J, 122 (14), Pang, Y. T.; Hazel, A. J.; Gumbart, J. C., "Uncovering the folding mechanism of pertactin: A comparative study of isolated and vectorial folding," pp. 2988–2995, Copyright (2023), with permission from Elsevier.

3.1 Introduction

Autotransporters are a class of proteins found in Gram-negative bacteria, primarily implicated in bacterial virulence. Historically, they were believed to possess all nec-

Y. T. (Andrew) Pang, *Biomolecules' Conformational Changes Studied by Simulations and Enhanced Sampling*, Springer Theses,
https://doi.org/10.1007/978-3-031-70602-8_3

essary functional elements for secreting their virulence factors—hence their name. However, recent studies suggest that this understanding may be incomplete [1–3]. Type Va autotransporters, in particular, consist of two principal domains: a C-terminal 12-stranded β-barrel ("β") domain [4–7] located in the outer membrane (OM), and an N-terminal β-helical ("passenger") domain [8], which passes through the β-barrel and carries the virulence factor itself [9]. The secretion mechanism of the passenger domain remains a subject of debate. It is known to proceed unidirectionally from the C- to N-terminus through the β-domain [10, 11] and hypothesized that secretion occurs via a vectorial folding pathway, where small β-structures formed post-secretion at the C-terminus prevent the peptide strand from slipping back through the β-domain [8, 11–13]. Contrarily, other studies have shown that folding of the passenger domain may not be necessary for secretion and highlighted the role of the β-barrel assembly machinery (BAM)) complex in orchestrating the membrane insertion of the β-domain and the concerted secretion process [10, 14–16]. Given the OM lacks ATP or other traditional energy sources, the folding of the passenger domain was once considered the primary energy source for secretion. Nonetheless, subsequent analyses suggest that the energy from folding contributes only a minor fraction to the total energy required for secretion [17].

The autotransporter pertactin from *Bordetella pertussis*, the causative agent of whooping cough, serves as a widely studied model for passenger-domain secretion. In vitro studies have demonstrated that the pertactin passenger domain takes several hours to fold and several days to reach equilibrium [8] (Fig. 3.1). This extended folding duration contrasts sharply with the rapid growth rates in vivo, where pathogenic bacteria typically have a doubling time of approximately 30–60 min. Additionally, experiments indicate that when OM secretion is stalled, the pertactin passenger domain remains unfolded in the periplasm for over 90 min, suggesting a coupling between secretion and folding processes in vivo [18]. A partly folded intermediate state consisting of the first six and a half C-terminal rungs of the β-helix, called the C-terminal core, was observed to be stable on its own and was initially thought be a scaffold for further folding in vivo [8, 19]. However, subsequent studies revealed that folding from this intermediate to the fully folded structure occurs more slowly than from the unfolded state, reclassifying the C-terminal core as an off-pathway intermediate rather than a folding scaffold [20]. Nevertheless, the stability of this core is essential for the effective secretion of pertactin [11], with the N-terminal half unable to fold in the absence of the stable C-terminal half [13]. Alternative mechanisms suggest that an extracellular loop on the β-domain may act as a scaffold [21, 22], and a hydrophobic cavity within the extracellular pore of the β-domain could facilitate the folding of the passenger domain [21]. Collectively, these findings highlight that pertactin follows distinctly different folding pathways during secretion compared to in vitro conditions, with the cellular environment significantly influencing the chosen pathway.

Kinetic models have posited that the folding of the passenger domain serves as the primary driving force for secretion efficiency [18]. Research has indicated that destabilizing the N-terminus of the passenger domain can enhance secretion efficiency, whereas destabilizing the C-terminus results in reduced efficiency [12, 13, 23]. Additionally, it has been observed that the N-terminus remains unfolded in

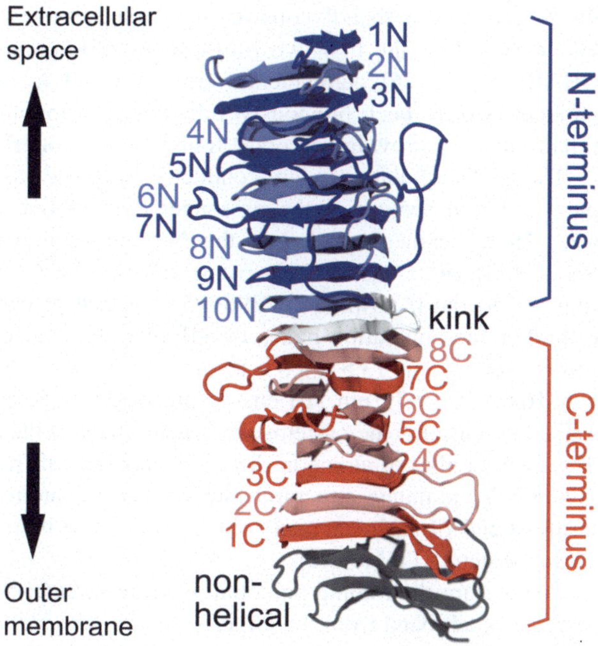

Fig. 3.1 Crystal structure of the pertactin passenger domain from *B. pertussis* (PDB ID: 1DAB). The structure is colored and labeled by rungs, with each band representing two consecutive rungs, named after the rung closest to the termini (e.g., rungs 1C and 2C form band 1C). Reprinted from Biophys J, 122 (14), Pang, Y. T.; Hazel, A. J.; Gumbart, J. C., "Uncovering the folding mechanism of pertactin: A comparative study of isolated and vectorial folding," pp. 2988–2995, Copyright (2023), with permission from Elsevier

the absence of the C-terminus, further underscoring the interdependence of these regions [24]. These findings suggest that a free energy gradient along the passenger domain β-helix is crucial for efficient secretion. This gradient is hypothesized to fulfill several roles: (1) preventing premature folding during secretion across the inner membrane into the periplasm, which occurs in the opposite direction to OM secretion, (2) delaying the formation of the native state while the protein resides in the periplasm, and (3) facilitating the more rapid, vectorial secretion process across OM into the extracellular space.

In this study, we employ steered molecular dynamics (SMD) simulations to gain insight into the folding process of the pertactin passenger domain by unfolding it. This approach allows us to probe the intricate dynamics and stability of the protein structure. Furthermore, we utilize enhanced sampling methods to juxtapose the energetics of the in vitro and in vivo folding pathways of the pertactin passenger domain. This comparative analysis aims to evaluate the plausibility and efficiency of vectorial folding as the primary energy source driving secretion.

3.2 Methods

3.2.1 System Construction

The structural model of the pertactin β-helical passenger domain was sourced from the Protein Data Bank (PDB), specifically PDB ID: 1DAB [25]. Throughout this chapter, a "rung" is defined as 2–3 connecting β-strands that complete a turn within the β-helix, and a "band" consists of two successive "rungs". For example, band 1C encompasses the first two rungs at the C-terminus of the β-helix (Fig. 3.1). CHARMM22* [26] parameters were employed to describe the protein. These parameters have been previously validated to accurately represent the folding of β-hairpins [27], providing a robust foundation for our simulations.

The SMD simulations were conducted in four distinct segments, each focusing on a different section of the passenger domain due to the protein's substantial size. These segments included: the N-terminus to rung 10N (D1–G284), rung 8N to rung 5C (A184–G391), rung 10N to the C-terminus (Y262–P539), and rung 4C to the C-terminus (V396–P539). Each protein segment was solvated in orthorhombic simulation boxes of TIP3P water molecules [28], with dimensions respectively set at $13.5 \times 8.0 \times 8.0$ nm, $13.5 \times 8.0 \times 8.0$ nm, $15.0 \times 7.0 \times 7.0$ nm, and $16.8 \times 7.0 \times 7.0$ nm. To ensure electrostatic neutrality and simulate physiological conditions, the systems were neutralized with a 0.15 M solution of NaCl. The total number of atoms in these systems ranged from approximately 70,000 to 106,000. To maintain structural integrity, all C_α atoms of the first band at the C-terminal end of each segment were restrained to their positions as defined in the crystal structure (Table B.1).

To measure the folding potential of mean forces (PMFs) for isolated bands, we extracted each band from the larger structure, as specified in Table B.2 which lists the residues involved for each system. Subsequently, each peptide was solvated in a $15.0 \times 7.0 \times 7.0$ nm box containing ~23,000 TIP3P water molecules [28] and neutralized with a 0.15 M NaCl solution, resulting in a total system size of ~70,000 atoms. To ensure proper alignment for simulation, we constrained the end-to-end vector of the peptide along the x-axis. This was achieved using the distanceXY function from the collective variables (colvars) module of NAMD [29], which maintained the distance between the first and last residues of the peptide along the y- and z-axes at zero. The force constant for this restraint was set to 110 kcal/mol·Å^2, ensuring stability and precise orientation throughout the simulation.

For assessing the vectorial folding PMFs of each band, we designed systems that included all residues C-terminal to the band in question. The construction ceased at residue 280 for band 1N and at residue 539 for band 2C. To accommodate the peptide in both its folded state and a fully unfolded and extended N-terminal rung, the water boxes were dimensioned accordingly, with the y- and z-dimensions ranging from ~6.5 to 8.0 nm and x-dimension stretching between ~11.7 and 19.3 nm. Each system was also neutralized with a 0.15 M NaCl solution. The total system sizes varied between ~53,000 and 106,000 atoms. To maintain structural integrity during simulations, the heavy backbone atoms of all β-sheet residues, excluding those in the first N-terminal rung, were restrained to their native conformations.

3.2.2 *Molecular Dynamics Simulations*

Molecular dynamics (MD) simulations were performed using NAMD 2.10–12 [30]. The system temperature was maintained at 310 K through Langevin dynamics, and the pressure was kept constant at 1 atm using the Langevin piston method [31]. The equations of motion were integrated using the RESPA multiple time-stepping algorithm, which allowed for a time step of 2 fs for both bonded interactions and short-range non-bonded interactions, and 4 fs for long-range electrostatic interactions. Long-range electrostatic forces were calculated using the particle-mesh Ewald (PME) method, implementing a real-space cutoff of 1.2 nm [32]. Short-range non-bonded Lennard-Jones interactions were truncated at 1.2 nm, with a potential switching function initiated at 1.0 nm to smoothly transition the potential energy to zero at the cutoff distance. To ensure structural integrity, bonds involving hydrogen atoms were constrained at their equilibrium lengths using the SETTLE algorithm for water molecules [33] and the SHAKE algorithm for all other molecules [34].

3.2.3 Steered Molecular Dynamics

SMD was employed to methodically unfold the passenger domain by applying a controlled force to the first C_α atom at the N-terminal end of each system segment. The force was exerted at a rate of 0.1 nm/ns in the $+x$-direction, with a force constant of 0.05 kcal/mol·nm^2. Prior to the application of force, the passenger domain was aligned along the x-axis to ensure that the pulling force correctly pull the N-terminal C_α away from the C-terminus, facilitating the intended unfolding process. The simulation was repeated three times for each segment to validate the consistency and reliability of the results. To prevent any interaction between the extended portion of the protein with its periodic images, the protein was trimmed periodically every 40–100 ns, depending on the extension observed during the simulation. Each segment was subjected to a different duration of applied force based on its length: the N-terminus to rung 10N for 540–560 ns, rung 8N to rung 5C for 220 ns, rung 10N to the C-terminus for 400 ns, and rung 4C to the C-terminus for 260 ns. The total distances pulled for each run varied slightly due to differences in the length of the trimmed parts.

3.2.4 Self-learning Adaptive Umbrella Sampling Procedure

We computed two-dimensional folding free energy landscapes for the β-helical bands of pertactin using umbrella sampling (US). Our choice of reaction coordinates was inspired by those previously utilized for the β-hairpin peptide, GB1 [27]. These coordinates include the number of native hydrogen bonds along the β-sheets (N_{hb}) and the radius of gyration of all internal hydrophobic side chains (R_{G}). The N_{hb} coordinate is determined by summing individual hydrogen bond collective variables, as defined in Eq. (3.1).

$$\text{hbf(donor, acceptor)} = \frac{1 - (|x_{\text{don}} - x_{\text{acc}}|/d_0)^6}{1 - (|x_{\text{don}} - x_{\text{acc}}|/d_0)^8}, \tag{3.1}$$

This hbf function is a smooth function from 0 to 1 where hbf$\approx$1 if two atoms are well within the cutoff (d_0 =0.33 nm), and hbf$\approx$0 when they are well outside the cutoff. Windows were spaced by 0.25 along N_{hb}, with a harmonic force constant of 25.0 kcal/mol, and by 0.05 nm along R_{G}, with a harmonic force constant of 0.625 kcal/mol·nm^2. All US simulations were run at 310 K. PMFs were generated via the weighted histogram analysis method (WHAM) [35, 36].

To initiate the self-learning adaptive umbrella sampling (SLUS) procedure [37, 38] for the isolated bands, we began with a preparatory phase consisting of a 10-ns MD simulation at 310 K, followed by another 10-ns simulation at an elevated temperature of 500 K. These simulations served to generate a diverse set of initial states for a 2D umbrella sampling campaign. For the SLUS procedure, we used a 5 kcal/mol free energy threshold, with each window in the simulation running for 2 ns per iteration. The initial set of windows (iteration 0) utilized replica exchange umbrella sampling (REUS) [39], while standard US was applied for all subsequent iterations. Upon completing the SLUS iterations, we extended the analysis by conducting an additional 10 ns/window of REUS across all the windows that had been generated during the procedure. Table B.2 details the number of SLUS iterations undertaken and the number of windows generated for each band.

For the vectorial folding PMFs, we employed metadynamics simulations [40] spanning ~50–150 ns for each system to generate extensive sampling windows—333 for band 1N and 210 for band 2C. Following the metadynamics phase, an initial round of 7 ns per window of REUS simulations was conducted for both systems. We utilized data from the final 5 ns of each window to compute the initial PMFs. The initial coverage of reaction coordinates provided by the metadynamics simulations was comprehensive enough that subsequent SLUS did not contribute additional exploration within the reaction coordinate space. To check convergence for band 2C, we ran an additional 5 ns/window of REUS simulations, using the final 10 ns/window from all the trajectories to calculate the PMF for this system.

3.3 Results

3.3.1 *The C-terminus Possesses a Stable Hydrophobic Core*

To probe the folding dynamics of the pertactin passenger domain, we employed steered molecular dynamics (SMD) simulations to unfold the protein. A pulling force of 0.05 kcal/mol·nm^2 was consistently applied to the N-terminal C_α atom, drawing it away from the C-terminus at a speed of 0.1 nm/ns. To ensure stability and focus the force's effect, all C_α atoms of the C-terminal band were fixed in place. This setup was replicated across three separate trials to ensure reliability of the results. Given the substantial size of the passenger domain, the unfolding procedure was segmented into four distinct runs, each featuring only a portion of the entire protein (see Sect. 3.2). To maintain the system within the confines of the simulation box and prevent interactions with its periodic images, the extended portions of the protein were periodically removed. Force spikes observed during the simulations were meticulously correlated with specific unfolding events (see Sect. B.1). Below, we delve into a few select unfolding events that illustrate significant aspects of the unfolding process.

In all three replicas of the SMD simulations, the pertactin passenger domain generally unfolded sequentially, strand-by-strand from the N-terminus to the C-terminus, without explicit restraints dictating the order of unfolding. Figure 3.2a depicts the resistance force profile observed during the unfolding process. Notably, each force spike corresponds to the detachment of a β-strand from the β-helix. Among these, the force spike occurring at a pulled distance of approximately 130 nm was particularly significant, as it displayed the highest peak force and the largest area under the force-extension (F-x) curve, consistently across all three

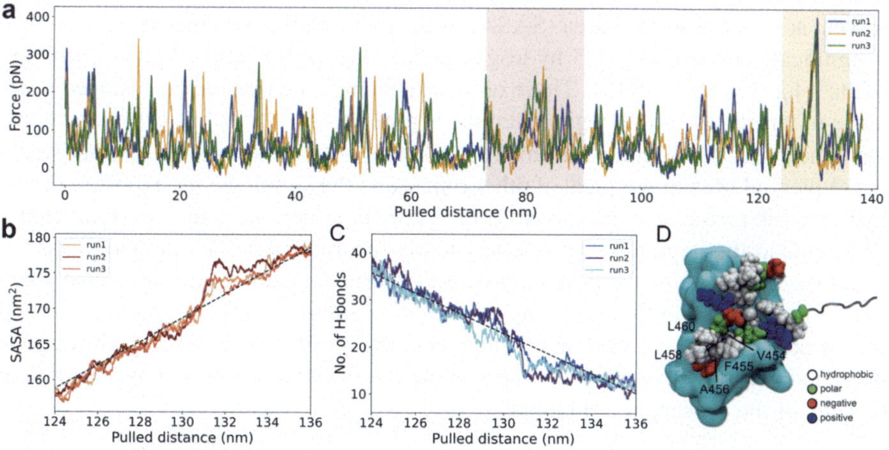

Fig. 3.2 SMD simulation results of the unfolding of the pertactin passenger domain. (**a**) Force-distance profiles for all three replicas of the simulation. The rungs of the β-helix unfolds one by one from C- to N-terminus throughout the simulation except when unfolding the kink (pale red, see Fig. 3.3). The force spike observed near the end of the simulation at a pulled distance of ~130 nm (pale orange) is consistently observed across all replicas and corresponds to the unfolding of the β-hairpin at V454–S461. (**b**) SASA of the hydrophobic residues and (**c**) number of hydrogen bonds against the pulled distance over the pale orange shaded region. A sharp increase in SASA and decrease in hydrogen bonds are observed at the same point as the force spike in (**a**). (**d**) Snapshot from the SMD simulations before the β-hairpin unfolds. The residues of band 1C are colored by their residue types, with hydrophobic, polar, positively-charged, and negatively-charged residues colored as white, green, red, and blue, respectively. Reprinted from Biophys J, 122 (14), Pang, Y. T.; Hazel, A. J.; Gumbart, J. C., "Uncovering the folding mechanism of pertactin: A comparative study of isolated and vectorial folding," pp. 2988–2995, Copyright (2023), with permission from Elsevier

simulations. This prominent spike was associated with the unfolding of the β-hairpin at V454–S461, which links the first two rungs at the C-terminus (rungs 1C and 2C, or band 1C) and interfaces with the non-β-helical C-terminal region. This β-hairpin, uniquely structured and comprising predominantly hydrophobic residues (Fig. 3.2d), is the sole connecting loop between the rungs that adopts a structured form. It is flanked by two unstructured loops in the folded state, which shield it from the aqueous environment. Previous research has highlighted that the C-terminus of the pertactin passenger domain forms a stable hydrophobic core resistant to Proteinase K (proK) digestion [19]. Our observations corroborate these experimental results, identifying the V454–S461 β-hairpin as the central element of this hydrophobic core. Moreover, the stability of the C-terminus is not exclusive to pertactin but is a recurring feature in many Type Va autotransporters, such as Pet [19]. These findings lend further support to the vectorial folding hypothesis, which posits a structured and sequential folding mechanism as integral to autotransporter secretion [19, 24].

Further analysis of the SMD simulations was conducted to elucidate the causes behind the substantial force spike observed during the unfolding of band 1C. Our examination pinpointed that this force spike coincided with a marked increase in the solvent accessible surface area (SASA) of the hydrophobic residues (Fig. 3.2b) and a significant reduction of ~5 hydrogen bonds (Fig. 3.2c) following the unfolding of the β-hairpin V454–S461. These observations indicate that the pronounced resistance to unfolding in this region stemmed from two primary physical interactions: the concurrent disruption of all hydrogen bonds within the hairpin, as opposed to the sequential breaking typical of other rungs, and the simultaneous exposure of five hydrophobic residues to the solvent (Fig. 3.2d). Together, these factors significantly contribute to the large force spike consistently recorded across all three replicas.

At the other end of the β-helix, we also observed a consistent force spike when unfolding the first and second rungs at the N-terminus (rungs 1N and 2N, or band 1N) at pulled distances of 0 and 5 nm, respectively (Fig. 3.2a). The spike suggests that band 1N may play a role in capping the β-helix and preventing aggregation at this end of the passenger domain [41].

3.3.2 A Kinked Rung Resists Unfolding

SMD simulation results from all three replicas consistently show that the pertactin passenger domain unfolds strand-by-strand. A notable exception is observed when the kink, spanning residues I277 to R291 and located centrally within the passenger domain, is engaged. This kink induces a subtle orientation change of approximately 20° between the C- and N-terminal halves, impacting the unfolding process at a pulled distance of ~80 nm. In the first and second runs, the β-strand adjacent to the kink from the following rung (A304–G314) disengages from the main structure and unfolds concurrently with the kink (Fig. 3.3a). The third run shows a more extensive detachment, resulting in a rupture between rungs 5C (P378–G391) and

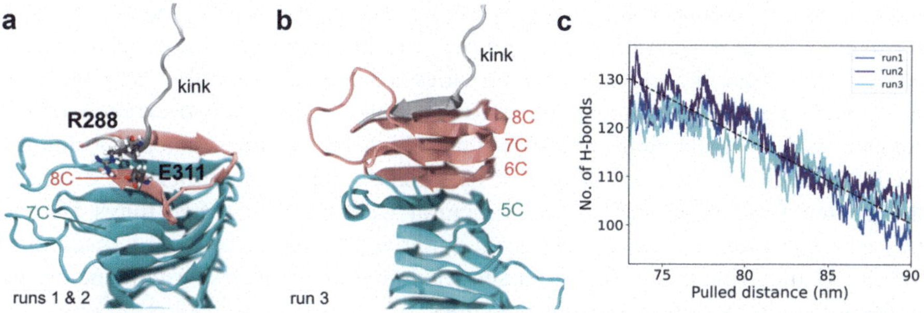

Fig. 3.3 Unique unfolding mechanism when the kink is being pulled off. (**a**, **b**) The pertactin passenger domain unfolds strand-by-strand throughout the SMD simulation except when the kink (I277–R291, silver) unfolds. (**a**) In runs 1 and 2, an adjacent β-strand (A304–G314, pink) detaches along with the kink, partly due to the strong electrostatic interactions between the positively charged residue R288 on the kink and the negatively charged residue E311. (**b**) In run 3, an even larger portion of the protein (rung 6C–8C, pink) detaches. (**c**) The events are accompanied by a drop of ∼8 hydrogen bonds, explaining the strong resistance force as seen in Fig. 3.2a. Reprinted from Biophys J, 122 (14), Pang, Y. T.; Hazel, A. J.; Gumbart, J. C., "Uncovering the folding mechanism of pertactin: A comparative study of isolated and vectorial folding," pp. 2988–2995, Copyright (2023), with permission from Elsevier

6C (P348–V364), further illustrating the structural vulnerability introduced by the kink (Fig. 3.3b). During these events, all three runs exhibited a significant force spike, exceeding 200 pN in height and spanning ∼6 nm in width (Fig. 3.2a). This mechanical stress was accompanied by the loss of roughly 8 hydrogen bonds (Fig. 3.3c).

Several factors contribute to the observed multi-strand detachment during the unfolding process, notably the unique arrangement of charged residues around the kink. A key interaction involves the positively charged residue R288 at the kink, which forms strong electrostatic bonds with two negatively charged residues on adjacent rungs, D265 and E311 (Fig. B.1b). SASA calculations indicate that D265 and E311 are among the top four charged residues with minimal exposure to the solvent, enhancing the stability of these electrostatic interactions within the β-helix (Fig. B.1a). Notably, the interaction between R288 and E311 persisted even after the unfolding of their respective rungs in runs 1 and 2 (Fig. 3.3a). As the unfolding progressed, rung 8C remained attached to the β-helix to maintain contact with the kink until it became too distant, leading to the eventual loss of all interactions. Previous experimental studies have characterized the stable C-terminal structural core as A335–P539, which includes the passenger domain from the C-terminus up to rung 7C [8]. This region aligns with the rupture points observed in our SMD simulation runs 1 and 2, highlighting the structural significance of these interactions in the stability and unfolding dynamics of the pertactin passenger domain.

Another important factor influencing the multi-strand detachment, as observed in run 3, is the distinctive geometry of the passenger domain. The C-terminal half of the domain resembles the shape of a boot. At the top of this "boot", the kink forms a larger cross-sectional area compared to the narrower "shaft" of the rungs below. This configuration not only provides a larger hydrophobic core but also facilitates an increased number of hydrogen bonds among the backbones of the rungs, thereby enhancing the overall structural integrity of the segment. During the SMD simulation, as tension is applied to the kink, the "shaft" efficiently channels the pulling force downward to the structural weak point between rungs 5C and 6C, which possess a smaller cross-sectional area. This efficient force transmission highlights the critical role of the passenger domain's geometry in influencing the unfolding dynamics under mechanical stress.

Another important factor for the multi-strand detachment, highlighted by run 3, is the geometry of the passenger domain, which has its C-terminal half roughly resembling the shape of a boot. The kink (the top of the "boot") has a larger cross-sectional area than the rungs down the "shaft", providing a larger hydrophobic core and allowing more hydrogen bonds between the backbones of the rungs, both strengthening its structural integrity. As the kink is being pulled, the shaft effectively transfers the pulling force to the structural weak point of the boot, disrupting the structure at a point with a small cross-sectional area, i.e., between rungs 5C and 6C.

Overall, our results suggest that the kink is more stable than the bands composed of rungs 5C and 6C as well as those composed of rungs 7C and 8C, leading to these two bands breaking before the kink unfolds. The enhanced stability of the kink could serve a functional benefit, particularly in facilitating the secretion of the long loop connecting rungs 9N and 10N (R226–W261), which includes the RGD motif crucial for cell adhesion [42] (Fig. 3.1). This structural arrangement not only underscores the mechanical resilience of certain domain regions but also aligns with the biological functionality of facilitating adhesion during the infection process.

3.3.3 Instability of Isolated Bands Revealed Through Potentials of Mean Force

To investigate the stability and folding dynamics of individual bands within the pertactin β-helical passenger domain, we utilized SLUS as outlined in Sect. 3.2 [37]. This approach was specifically applied to the first three bands at both the C- and N-termini, as well as the centrally located kinked band, which includes two complete rungs and one incomplete rung (detailed in Table B.2 and Fig. 3.1). The chosen reaction coordinates for this study were consistent with those previously applied to the GB1 β-hairpin [27]. Following the SLUS procedure, an extensive series of REUS simulations were conducted, with each window running for 15 ns. This rigorous analysis generated detailed 2D-folding PMFs, which are illustrated in Fig. 3.4a for the terminal bands and in Fig. B.2 for the kinked band.

The isolated state of each band proved to be unstable, particularly highlighted by bands 1C and 3N, where the native configuration was not retained during the

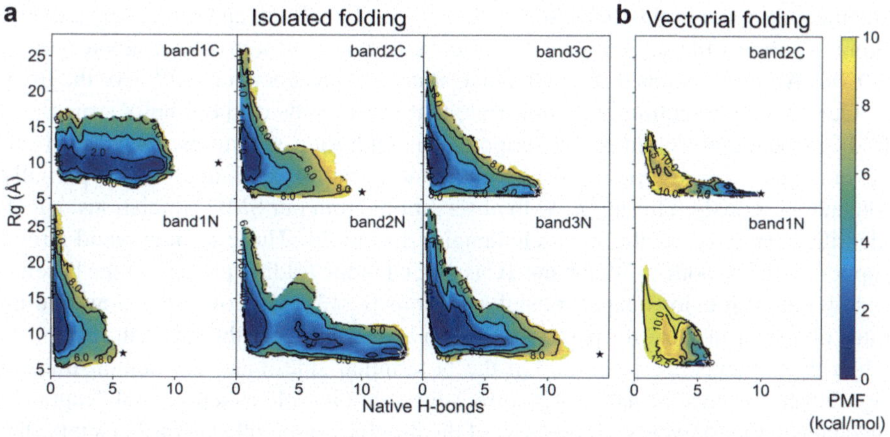

Fig. 3.4 PMFs for the folding of pertactin passenger domain bands. (**a**) 2D PMFs for isolated folding of bands 1–3C and 1–3N along two collective variables, N_{hb} and R_{G}, showing the native structures (indicated with a star) are unstable except for bands 1C and 2N. (**b**) 2D PMFs for vectorial folding of bands 2C and 1N along the same variables, showing that the native structures become highly favorable. Reprinted from Biophys J, 122 (14), Pang, Y. T.; Hazel, A. J.; Gumbart, J. C., "Uncovering the folding mechanism of pertactin: A comparative study of isolated and vectorial folding," pp. 2988–2995, Copyright (2023), with permission from Elsevier

SLUS procedure. These bands rapidly lost their native structures within the initial nanosecond of the equilibrium simulations conducted at 310 K, and the energy barriers required to refold them were insurmountable given the 5-kcal/mol cutoff set for the SLUS simulations. For the other bands examined, only bands 3C and 2N demonstrated a free energy minimum close to their native states. Interestingly, band 1C, while predominantly unfolding, did exhibit two minor energy minima at $\sim$3.5 and $\sim$8.5 hydrogen bonds, though all observed minima were separated by energy barriers of $<$2 kcal/mol. Similarly, band 2N displayed intermediate folded states between 5 and 8 hydrogen bonds, with minima also delineated by barriers $<$3 kcal/mol.

Among all the bands examined, band 1C demonstrated the highest stability. PMF analysis reveals that band 1C exists in three states, each occupying roughly equal proportions: a fully unfolded state and two partially folded states. When these partially folded states are combined, the folding free energy (ΔG_{fold}) is calculated to be -0.8 ± 0.5 kcal/mol, indicating that the band remains in a partially folded configuration $\sim$79% of the time. Focusing solely on the most structured minimum, which retains $\sim$60% of the native hydrogen bonds, the ΔG_{fold} shifts to $+0.2\pm0.4$ kcal/mol, with the time spent in this structured state reducing to $\sim$44%. Following band 1C, band 2N exhibits the second highest stability with three discernible minima: a fully unfolded state, a partially folded state ($\sim$40%–60% folded), and the fully folded native state. When the partially and fully folded states are combined, the ΔG_{fold} is $+0.7\pm0.6$ kcal/mol, with the native structure being present $\sim$25% of the time. Considering only the native-state minimum, the ΔG_{fold}

and the time spent in this state change to $+1.7\pm1.1$ kcal/mol and ~6%, respectively. Band 3C shows the next highest occurrence of any native structure, albeit only at ~0.7%. All other bands displayed native structural presence for <0.1% of the time.

Our PMFs compellingly demonstrate that most bands of the β-helix struggle to fold independently without the support of neighboring structures. However, band 1C emerges as an exception, displaying notable stability and an ability to partially fold autonomously. This aligns with observations from our SMD simulations, which identified band 1C as the most challenging to unfold. These findings lend strong support to the hypothesis that band 1C is crucial in the folding process of the β-helix, corroborating previous experimental assertions [8, 13]. Moreover, the slow folding rates observed in vitro for pertactin [8] can be attributed to the inherent instability when these bands are isolated. On the N-terminal side, band 2N demonstrates a significant presence of native structure, suggesting its role as a potential "capping" mechanism that prevents aggregation of the β-helix—a hypothesis further supported by our SMD findings and aligning with earlier studies [41].

3.3.4 *Vectorial Folding Stabilizes the β-Helical Bands*

To better understand the impact of vectorial folding on band stability, we calculated the free-energy landscapes for bands 2C and 1N by restraining all C-terminal β-strands to their native conformations, allowing only the most N-terminal rung within each band to move freely (see Sect. 3.2 for more details). The resulting PMFs are illustrated in Fig. 3.4b.

In sharp contrast to their instability when isolated (as shown in Fig. 3.4a), these bands demonstrate considerably higher stability, with PMF values around -10 kcal/mol, when subjected to vectorial folding. This arrangement simulates the sequential folding process that occurs in vivo, with each band stabilizing atop its adjacent C-terminal counterpart. The imposition of restraints on one of the two rungs in each band markedly reduces their configurational entropy, thereby elevating the free energy of the unfolded state and significantly enhancing the overall stability of the folded state.

In the vectorial folding hypothesis, as the peptide is secreted through the OM, only small segments immediately N-terminal to the already secreted and folded bands of the β-helix are exposed to the extracellular space. Meanwhile, the rest of the peptide remains dynamically active within the periplasmic side. This configuration likely results in even higher entropy for the periplasmic portion than observed in our truncated isolated systems. Consequently, the increase in free energy for the unfolded state in vivo could be substantially greater than what we have documented, further enhancing the stability of the folded structures on the extracellular side. The vectorial folding pathway not only prevents the backsliding of the peptide through the OM but also effectively eliminates the barrier to folding. This mechanism significantly stabilizes the folded state during the secretion process, providing a streamlined path for the peptide to adopt its functional conformation.

3.4 Discussion

This study was designed to deepen our understanding of the folding mechanisms of the pertactin passenger domain, a critical adhesion contributing to the pathogenicity of *B. pertussis*. Employing a suite of computational techniques, including SMD, SLUS, and REUS, we explored the stability and folding dynamics across various bands of the pertactin passenger domain. In our SMD simulations, we specifically manipulated the structure by anchoring the C-terminal band and applying tensile force at the N-terminal end. This approach revealed that the pertactin passenger domain predominantly unfolds in a strand-by-strand manner from the N- to the C-terminus, with notable resistance occurring at a centrally located kink. The SMD results indicated that band 1C exhibited the greatest resistance to unfolding, aligning with previous findings that highlight a stable C-terminal hydrophobic core [8, 19]. The most significant resistance was encountered when unfolding the β-hairpin (V454–S461) that bridges rungs 1C and 2C, situated above a hydrophobic segment of the non-β-helical C-terminus. The unfolding of this hairpin involved breaking all hydrogen bonds simultaneously and exposing a substantial number of hydrophobic residues to the solvent, which significantly contributed to its high resistance to unfolding.

Our SMD simulations have elucidated the unfolding behavior of the pertactin passenger domain, while our SLUS and REUS simulations have detailed the folding energetics of its individual bands. Analysis of the PMFs indicates that most bands struggle to fold independently, with band 1C being a notable exception. This band shows a negative ΔG_{fold} of -0.8 ± 0.5 kcal/mol, highlighting its energetic favorability even in isolation. The convergence of findings from both SMD and REUS reinforces the exceptional stability of band 1C, positioning it as a potential folding nucleus within the passenger domain. This stability suggests that band 1C could play a pivotal role in initiating or propagating the folding process, underscoring its importance in the overall structural assembly and function of the pertactin passenger domain.

While the folding of band 2N in isolation exhibits a positive ΔG_{fold} of $+0.7\pm0.6$ kcal/mol, indicating it is energetically unfavorable, this band still maintains a significant population ($\sim$25%) in its native structure. This suggests that, alongside band 1C, band 2N may play a critical role in the structural organization of the pertactin passenger domain. Our findings propose that these two terminal bands potentially serve as scaffolds, promoting the propagation of the β-helical structure inward towards the kinked region during folding, likely contributes to the concerted folding process observed in vitro for the pertactin passenger domain [20]. Furthermore, our simulations indicate that the vectorial folding pathway provides substantial stabilization for the native states of bands 2C and 1N. This stabilization is achieved by sequentially providing a structural template that facilitates the correct folding of each successive rung.

In conclusion, our study provides valuable insights into the folding mechanisms and stability of the pertactin passenger domain, a representative model for type

Va autotransporters. We have detailed how specific structural features of this protein contribute to the virulence of *B. pertussis*. These insights not only enhance our understanding of pathogenic mechanisms but also offer potential avenues for developing novel vaccines and therapeutics. Specifically, targeting the vectorial folding pathway to inhibit the folding and secretion of pertactin could curtail virulence while reducing the likelihood of developing resistance. Such strategies could be pivotal in the ongoing effort to combat pertussis more effectively.

References

1. A. Sauri et al., The Bam (Omp85) complex is involved in secretion of the autotransporter haemoglobin protease. Microbiology **155**(Pt 12), 3982–3991 (2009)
2. R. Wu et al., Plasticity within the barrel domain of BamA mediates a hybrid-barrel mechanism by BAM. Nat. Commun, **12**(1), 7131 (2021)
3. K.M. Kuo, D. Ryoo, K. Lundquist, J.C. Gumbart, Modeling intermediates of BamA folding an outer membrane protein. Biophys. J. **121**(17), 3242–3252 (2022)
4. C.J. Oomen, P. van Ulsen, P. van Gelder, M. Feijen, J. Tommassen, P. Gros, Structure of the translocator domain of a bacterial autotransporter. EMBO J. **23**(6), 1257–1266 (2004)
5. T.J. Barnard, N. Dautin, P. Lukacik, H.D. Bernstein, S.K. Buchanan, Autotransporter structure reveals intra-barrel cleavage followed by conformational changes. Nat. Struct. Mol. Biol. **14**(12), 1214–1220 (2007)
6. B. van den Berg, Crystal structure of a full-length autotransporter. J. Mol. Biol. **396**(3), 627–633 (2010)
7. I. Meuskens, A. Saragliadis, J.C. Leo, D. Linke, Type V secretion systems: an overview of passenger domain functions. Front. Microbiol. **10**, 1163 (2019)
8. M. Junker et al., Pertactin β-helix folding mechanism suggests common themes for the secretion and folding of autotransporter proteins. Proc. Natl. Acad. Sci. U. S. A. **103**(13), 4918–4923 (2006)
9. D.L. Leyton, A.E. Rossiter, I.R. Henderson, From self sufficiency to dependence: Mechanisms and factors important for autotransporter biogenesis. Nat. Rev. Microbiol. **10**, 213–225 (2012)
10. R. Ieva, H.D. Bernstein, Interaction of an autotransporter passenger domain with BamA during its translocation across the bacterial outer membrane. Proc. Natl. Acad. Sci. U. S. A. **106**(45), 19120–19125 (2009)
11. M. Junker, R. N. Besingi, P. L. Clark, Vectorial transport and folding of an autotransporter virulence protein during outer membrane secretion. Mol. Microbiol. **71**(5), 1323–1332 (2009)
12. J.H. Peterson, P. Tian, R. Ieva, N. Dautin, H.D. Bernstein, Secretion of a bacterial virulence factor is driven by the folding of a C-terminal segment. Proc. Natl. Acad. Sci. U. S. A. **107**(41), 17739–17744 (2010)
13. J.P. Renn, M. Junker, R.H. Besingi, E. Braselmann, P.L. Clark, ATP-independent control of autotransporter virulence protein transport via the folding properties of the secreted protein. Chem. Biol. **19**, 287–296 (2012)
14. R. Ieva, P. Tian, J.H. Peterson, H.D. Bernstein, Sequential and spatially restricted interactions of assembly factors with an autotransporter β domain. Proc. Natl. Acad. Sci. U. S. A. **108**(31), E343–E391 (2011)
15. O. Pavlova, J.H. Peterson, R. Ieva, H.D. Bernstein, Mechanistic link between β barrel assembly and the initiation of autotransporter secretion. Proc. Natl. Acad. Sci. U. S. A. **110**(10), E938–E947 (2013)
16. D. Ryoo, M.O. Rydmark, Y.T. Pang, K.P. Lundquist, D. Linke, J.C. Gumbart, BamA is required for autotransporter secretion. Biochim. Biophys. Acta Gen. Subj. **1864**, 129581 (2020)

17. W. Kang'ethe, H.D. Bernstein, Charge-dependent secretion of an intrinsicially disordered protein via the autotransporter pathway. Proc. Natl. Acad. Sci. U. S. A. **110**(45), E4246–E4255 (2013)
18. I. Drobnak, E. Braselmann, P.L. Clark, Multiple driving forces required for efficient secretion of autotransporter virulence proteins. J. Biol. Chem. **290**(16), 10104–10116 (2015)
19. J.P. Renn, P.L. Clark, A conserved stable core structure in the passenger domain β-helix of autotransporter virulence proteins. Biopolymers **89**(5), 420–427 (2008)
20. M. Junker, P.L. Clark, Slow formation of aggregation-resistant β-sheet folding intermediates. Proteins Struct. Func. Bioinf. **78**, 812–824 (2010)
21. Y. Zhai et al., Autotransporter passenger domain secretion requires a hydrophobic cavity at the extracellular entrance of the β-domain pore. Biochem. J. **435**, 577–587 (2011)
22. X. Yuan et al., Molecular basis for the folding of β-helical autotransporter passenger domains. Nat. Commun. **9**, 1395 (2018)
23. R.N. Besingi, J.L. Chaney, P.L. Clark, An alternative outer membrane secretion mechanism for an autotransporter lacking a C-terminal stable core. Mol. Microbiol. **90**(5), 1029–1045 (2013)
24. E. Braselmann, P.L. Clark, Autotransporters: The cellular environment reshapes a folding mechanism to promote protein transport. J. Phys. Chem. Lett. **3**, 1063–1071 (2012)
25. P. Emsley, I.G. Charles, N.F. Fairweather, N.W. Isaacs, Structure of *Bordetella pertussis* virulence factor P.69 pertactin. Nature **381**, 90–92 (1996)
26. S. Piana, K. Lindorff-Larsen, D.E. Shaw, How robust are protein folding simulations with respect to force field parametrization? Biophys. J. **100**, L47–L49 (2011)
27. A.J. Hazel, E.T. Walters, C.N. Rowley, J.C. Gumbart, Folding free energy landscapes of β-sheets with non-polarizable and polarizable CHARMM force fields. J. Chem. Phys. **149**, 072317 (2018)
28. W.L. Jorgensen, J. Chandrasekhar, J.D. Madura, R.W. Impey, M.L. Klein, Comparison of simple potential functions for simulating liquid water. J. Chem. Phys. **79**(2), 926–935 (1983)
29. G. Fiorin, M.L. Klein, J. Hénin, Using collective variables to drive molecular dynamics simulations. Mol. Phys. **111**(22–23), 3345–3362 (2013)
30. J.C. Phillips et al., Scalable molecular dynamics with NAMD. J. Comput. Chem. **26**(16), 1781–1802 (2005)
31. S.E. Feller, Y.H. Zhang, R.W. Pastor, B.R. Brooks, Constant pressure molecular dynamics simulations — The Langevin piston method. J. Chem. Phys. **103**, 4613–4621 (1995)
32. T. Darden, D. York, L. Pedersen, Particle mesh Ewald: An $N \cdot \log(N)$ method for Ewald sums in large systems. J. Chem. Phys. **98**(12), 10089–10092 (1993)
33. S. Miyamoto, P.A. Kollman, Settle: An analytical version of the SHAKE and RATTLE algorithm for rigid water models. J. Comput. Chem. **13**, 952–962 (1992)
34. J.-P. Ryckaert, G. Ciccotti, H.J.C. Berendsen, Numerical integration of the Cartesian equations of motion of a system with constraints: Molecular dynamics of n-alkanes. J. Comput. Phys. **23**, 327–341 (1977)
35. S. Kumar, J.M. Rosenberg, D. Bouzida, R.H. Swendsen, P.A. Kollman, The weighted histogram analysis method for free-energy calculations on biomolecules. I. The method. J. Comput. Chem. **13**(8), 1011–1021 (1992)
36. A. Grossfield, WHAM: The weighted histogram analysis method, version 2.0.6. http://membrane.urmc.rochester.edu/content/wham
37. W. Wojtas-Niziurski, Y. Meng, B. Roux, S. Bernèche, Self-learning adaptive umbrella sampling method for the determination of free energy landscapes in multiple dimensions. J. Chem. Theory Comput. **9**, 1885–1895 (2013)
38. A.J. Hazel et al., Conformational dynamics of AcrA govern multidrug efflux pump assembly. ACS Infect. Dis. **5**, 1926–1935 (2019)
39. Y. Sugita, A. Kitao, Y. Okamoto, Multidimensional replica-exchange method for free-energy calculations. J. Chem. Phys. **113**(15), 6042–6051 (2000)
40. A. Laio, M. Parrinello, Escaping free-energy minima. Proc. Natl. Acad. Sci. U. S. A. **99**(20), 12562–12566 (2002)

41. A.W. Bryan Jr, J.L. Starner-Kreinbrink, R. Hosur, P.L. Clark, B. Berger, Structure-based prediction reveals capping motics that inhibit β-helix aggregation. Proc. Natl. Acad. Sci. U. S. A. **108**(27), 11099–11104 (2011)
42. P. Everest et al., Role of the Bordetella pertussis P.69/pertactin protein and the P.69/pertactin RGD motif in the adherence to and invasion of mammalian cells. Microbiology **142**, 3261–3268 (1996)

Chapter 4
SARS-CoV-2 Spike Opening Dynamics and Energetics Reveal the Individual Roles of Glycans and Their Collective Impact

The trimeric spike (S) glycoprotein, which protrudes from the SARS-CoV-2 viral envelope, binds to human ACE2, initiated by at least one protomer's receptor binding domain (RBD) switching from a "down" (closed) to an "up" (open) state. Here, we used large-scale molecular dynamics simulations and two-dimensional replica exchange umbrella sampling calculations with more than a thousand windows and an aggregate total of 160 μs of simulation to investigate this transition with and without glycans. We find that the glycosylated spike has a higher barrier to opening and also energetically favors the down state over the up state. Analysis of the S-protein opening pathway reveals that glycans at N165 and N122 interfere with hydrogen bonds between the RBD and the N-terminal domain in the up state, while glycans at N165 and N343 can stabilize both the down and up states. Finally, we estimate how epitope exposure for several known antibodies changes along the opening path. We find that the BD-368-2 antibody's epitope is continuously exposed, explaining its high efficacy. Reproduced with permission from Pang, Y. T.; Acharya, A.; Lynch, D. L.; Pavlova, A.; Gumbart, J. C. "SARS-CoV-2 spike opening dynamics and energetics reveal the individual roles of glycans and their collective impact." *Commun Biol*, vol. 5, no. 1, pp. 1170, 2022.

4.1 Introduction

The COVID-19 pandemic, caused by the SARS-CoV-2 virus, rapidly escalated into a global crisis, significantly impacting public health and economies worldwide [1]. The development and widespread distribution of vaccines [2], monoclonal antibody therapies [3], and antiviral drugs [4] have been crucial in mitigating the virus's spread. However, the persistent threat from emerging variants, such as the Omicron sub-lineages [5], along with the potential for future coronavirus outbreaks [6],

Y. T. (Andrew) Pang, *Biomolecules' Conformational Changes Studied by Simulations and Enhanced Sampling*, Springer Theses,
https://doi.org/10.1007/978-3-031-70602-8_4

underscores the urgent need to deeply understand the viral life cycle. This includes mechanisms of recognition, binding, infection, and immune evasion.

The infection process of SARS-CoV-2 begins with the virus's recognition and attachment to the host cell's angiotensin-converting enzyme 2 (ACE2) receptor [7, 8]. This crucial step is facilitated by the spike (S) protein of SARS-CoV-2, a homotrimeric class I fusion glycoprotein that extends from the viral surface. The publication of the S-protein sequence early in 2020, along with prior structural studies on similar betacoronaviruses, enabled the swift determination of the solubilized, pre-fusion stabilized ectodomain structures of the S-protein [9–11]. These efforts also led to the elucidation of the full-length spike structures [12] and complete virion structures [13].

Each protomer of the SARS-CoV-2 spike protein is divided into S1 and S2 subunits, separated by a multibasic furin cleavage site. The S1 subunit houses the receptor binding domain (RBD), which is crucial for recognizing the host cell, while the S2 subunit includes the machinery necessary for membrane fusion and viral entry [7]. The spike protein acts as a principal antigenic target, with multiple epitopes such as the RBD and the N-terminal domain (NTD) recognized by the human immune system [14–17]. Recombinant RBD has been suggested as a potential inhibitor to block entry of SARS-CoV-2 [18]. Additionally, glycosylation of the S-protein plays a role in evading the host immune system by shielding the virus [19–21].

The spike protein undergoes transitions between down and up conformational states via a hinge-like movement, exposing the receptor binding motif (RBM), which consists of RBD residues from S438 to Q506 [22]. In the down state, the RBM remains hidden within the inter-protomer interface, necessitating sporadic shifts to the up state to facilitate ACE2 binding.

Cryo-electron microscopy (cryo-EM) studies have yielded detailed structural insights into the 'up' and 'down' conformational states of the spike protein [23]. Despite these advancements, the dynamics of these states and their transitions remain underexplored. Single-molecule fluorescence resonance energy transfer (FRET) has been used to illustrate the stochastic nature of these transitions, occurring over timescales from milliseconds to seconds [24]. Notably, additional single-molecule FRET studies have identified changes in the kinetics of down-up transitions in mutant forms of the spike protein [25]. Molecular dynamics (MD) simulations complement these experimental approaches by providing atomic-level detail of the intermediate states between the down and up conformations, critical for understanding the dynamics of spike protein opening. These simulations have elucidated aspects such as structural stability, the influence of glycosylation on conformational states, interactions between residues, and the specifics of ACE2 binding [20, 21, 26–29].

Steered MD and targeted MD simulations have elucidated the pathways for S-protein opening, providing key insights into the mechanics of this process [29–31]. Additionally, extensive use of advanced sampling techniques, such as weighted ensemble [32] and fluctuation amplification of specific traits (FAST) adaptive sampling combined with Folding@home [33], has revealed diverse pathways for the opening of the S-protein. Moreover, the emerging details of the energy landscape associated with these conformational transitions, essential for viral binding and entry, are increasingly being understood [27, 30, 31, 34]. This understanding is enhanced by integrating cryo-EM data with MD simulations, which have identified multiple sub-populations within both the down and up states of the S-protein [35, 36].

A recent study by Amaro and colleagues emphasized the functional significance of glycans at N165 and N234, demonstrating roles beyond mere viral shielding [26]. Separate equilibrium simulations of the S-protein's down and up states revealed that upon the RBD transitioning to the up state, the glycan at N234 rotates into the newly created void, thus stabilizing this conformation.

Furthermore, MD simulations and mutagenesis studies have uncovered the role of the glycan at N343 in the dynamics of RBD opening and ACE2 binding [32]. The findings suggest that the N343 glycan facilitates the conformational transition by "lifting" the RBD through sequential interactions with various RBD residues, a process described as "glycan gating."

In this study, we describe newly determined 2D free-energy landscapes for the opening and closing transitions of the SARS-CoV-2 S-protein, utilizing replica exchange umbrella sampling (REUS) simulations conducted on the pre-exascale supercomputer Summit at Oak Ridge National Lab (ORNL). These simulations, executed for both glycosylated and un-glycosylated forms of the S-protein, explore the impact of glycans on each state and the kinetics of spike protein opening. Additionally, we analyze the exposure of key epitopes on the S-protein surface, offering a dynamic perspective on antibody binding along the spike-opening pathway. Finally, results from equilibrium MD simulations for both glycosylated and un-glycosylated systems in the down and up states are reported to further clarify the stabilizing influence of glycans.

4.2 Methods

4.2.1 Model Building

We utilized cryo-EM structures from Walls et al. [10] (Protein Data Bank (PDB): 6VYB) and Cai et al. [12] (PDB: 6XR8) to model the up and down states of the S-protein using VMD [37]. The down-state structure provided by Cai et al. is a detergent-purified full-length wild-type (WT) S-protein construct resolved at 2.9 Å [12], notable for (1) a more defined NTD featuring a novel glycosylation site at N17, (2) retention of the WT sequence in the central helix/loop region rather than the pre-fusion stabilizing 2PP mutation, and (3) an approximately 25-residue segment (residues 828–853) previously unresolved near a key lysine (K854), which forms a salt bridge with D614. The disruption of this bridge has been associated with the heightened infectivity of the D614G SARS-CoV-2 strains.

Additionally, Cai et al. [12] discovered two previously unreported disulfide bonds: one in the N-terminal (C15-C136) and another in the central helix/loop region (C840-C851). We incorporated these missing residues in the down-state structures using SWISS-MODEL [38], with the 6XR8 structure serving as a template. The model was cleaved at the furin cleavage site between residues R685 and S686. The S1 subunit, comprising the RBD and NTD, appears more compact in the 6XR8 structure compared to 6VXX, while the S2 subunit, containing the HR1 and CH, aligns between them [12].

We included ten disulfide linkages in the S1 subunit between residues C15-C136, C131-C166, C291-C301, C336-C361, C379-C432, C391-C525, C480-C488, C538-C590, C617-C649, and C662-C671, and five in the S2 subunit: C738-C760, C743-C749, C840-C851, C1032-C1043, and C1082-C1126. Missing segments in the up state were constructed using the minimized down-state model. Glycosylation sites at N17, N61, N74, N122, N149, N165, N234, N282, T323, N331, N343, N603, N616, N657, N709, N717, N801, N1074, N1098, and N1134 were included, each protomer having 19 N-linked and 1 O-linked glycan, making up a total of 60 glycans per S-protein trimer. Glycans were modeled using the most populous species identified in mass spectrometry studies by Crispin et al. [19] and added via the GLYCAM Web server (https://glycam.org) [39, 40].

All systems had missing hydrogen atoms added and were solvated in a 195 $\times$ 193 $\times$ 212 Å^3 water box. Na^+ and Cl^- ions were adjusted to achieve a salt concentration of approximately 0.150 M. For REUS calculations, consistency in the number of atoms across protein, water, glycan, and ion components was maintained. The systems (up or down states) contained a total of 758,531 atoms (63,312 protein and glycan atoms, 661 Na^+, 652 Cl^-) with glycans, and 633,864 atoms (52,476 protein atoms, 549 Na^+, 546 Cl^-) without glycans. The variance in ion numbers results from the larger water box required to accommodate the glycans and the negative charge of the sialic acid (Neu5Ac in Fig. C.13) present in the O-linked glycans.

4.2.2 Molecular Dynamics (MD) Simulations

All simulations were carried out using NAMD 2.14 [41, 42], employing the CHARMM36m protein force field [43], the CHARMM36 glycan force field [44], and TIP3P water model [45]. Systems were equilibrated over 5 ns at a constant temperature of 310 K and pressure of 1 atm using Langevin dynamics and piston [46]. Simulations operated with a 4-fs time step, facilitated by hydrogen mass repartitioning (HMR) [47, 48]. Long-range electrostatics were computed every time step via the particle-mesh Ewald (PME) method [49], and a short-range cutoff for Lennard-Jones interactions was established at 12 Å, with a switching function initiating at 10 Å. Bonds involving hydrogen were constrained using the SETTLE algorithm for water molecules and the SHAKE algorithm for all other molecules.

Metadynamics Simulations

Post-equilibration, two independent metadynamics simulations were conducted for each system (glycosylated WT, un-glycosylated WT, and un-glycosylated with a diproline mutation), originating either from the down or up state. We used the colvars module of NAMD to define all collective variables [50]. Biases were applied with a hill height of 0.2 kcal/mol and a width of 1.0 Å and 1.0° for d and ϕ, respectively, at a deposition rate of 1 ps^{-1}. The ranges for d and ϕ were confined within the structural limits of the down (PDB: 6XR8) and up (PDB: 6VYB) states from cryo-EM studies. Temperature control was maintained at 310 K under constant volume conditions using Langevin dynamics.

REUS Simulation Seeding

The trajectories from metadynamics served as seeds for REUS simulations. Initially, two potential of mean forces (PMFs) were generated for each system, one from down-state and another from up-state metadynamics. The conformational space was segmented into small grids, each designated as a separate REUS window in subsequent steps. Windows with PMFs falling below a specified threshold from both down- and up-state metadynamics (Table C.2) were excluded from further REUS simulations. For the retained grids, a snapshot was selected from either the down- (F_{down}) or up-state (F_{up}) metadynamics trajectory, based on the Boltzmann weight of their respective PMF.

$$P_{\text{down/up}}(d, \phi) = \frac{\exp\left[-\beta F_{\text{down/up}}(d, \phi)\right]}{\exp[-\beta F_{\text{down}}(d, \phi)] + \exp\left[-\beta F_{\text{up}}(d, \phi)\right]} \tag{4.1}$$

If $P_{\text{down}} > P_{\text{up}}$, a snapshot from the down-state metadynamics trajectory was selected for that grid, and conversely for grids where $P_{\text{up}} > P_{\text{down}}$. This approach led to 387 windows for the glycosylated WT system (181 from down, 206 from up), 392 windows for the un-glycosylated WT system (79 from down, 313 from up), and 353 windows for the un-glycosylated system with the diproline mutation (119 from down, 234 from up).

Simulated Annealing of Glycans

For the glycosylated systems, simulated annealing was applied to all glycans in snapshots obtained from metadynamics simulations. Starting at a temperature of 1000 K, the systems were methodically cooled to physiological temperature in steps at 823, 677, 557, 458, 377 K, and finally 310 K. Each temperature step was maintained for a duration of 2 ns. Throughout these simulations, all protein atoms were constrained, allowing only the glycans, water molecules, and ions to move freely. Due to the elevated atomic velocities at temperatures of 1000 and 823 K, a reduced time step of 2 fs was employed, along with HMR. For temperatures 677 K and below, the time step was extended to 4 fs, continuing with the use of HMR. The simulation parameters otherwise mirrored those employed in earlier stages. The final conformations of the glycans were verified against expected structures, confirming their consistency (Fig. C.14).

REUS Simulations

With the initial configurations set, REUS simulations were conducted for each system along the collective variables d and ϕ, defined within specific ranges. These ranges spanned from 44.0 to 72.5 Å for d and from −56.0° to 1.0° for ϕ. The restraining force constants for d were adjusted to ensure adequate overlap and exchange between neighboring windows, while those for ϕ were held constant at 1.0 kcal/mol/deg^{-2}. The durations of the REUS simulations were 32, 36, and 29 ns for the WT glycosylated, WT un-glycosylated, and un-glycosylated systems with diproline mutation, respectively. The simulation parameters mirrored those used in the metadynamics simulations. The PMF for each system was computed using Multistate Bennett Acceptance Ratio (MBAR), implemented in pymbar [51]. For the WT glycosylated and un-glycosylated systems, three additional rounds of REUS simulations were conducted, using snapshots extracted in a similar manner to those from the metadynamics simulations.

In the second round, simulation regions were re-selected using a new energy cutoff based on the PMFs computed from the initial REUS round. The restraining force constants and window centers were recalibrated. The second round of REUS for the WT glycosylated and un-glycosylated systems lasted 37 and 56 ns, respectively.

Based on the results from the second round of REUS, a third round was initiated, expanding the PMF regions to fully encompass the identified energy wells. This round involved two independent REUS simulations for each of the energy wells, with ranges for d and ϕ set from 40.5 to 51.0 Å and −66.5° to −21.5° for the down-state, and 57.5 to 99.5 Å and −35.5° to 45.5° for the up-state, respectively. New metadynamics simulations were conducted to seed regions not previously sampled by REUS. The third round lasted 16 and 12 ns for the WT glycosylated down- and up-state systems, and 33 and 28 ns for the un-glycosylated systems, respectively.

A final round of REUS was conducted to span the full range from down to up state, integrating all windows from rounds 2 and 3, with additional ones on the edges to ensure coverage of all low-energy regions. The final round for the WT un-glycosylated system lasted 36 ns, directly utilizing the last frames from previous rounds. For the glycosylated system, new seeding structures were generated from the un-glycosylated REUS for windows around the energy barrier, bridging the structural gap in the glycosylated REUS. The glycosylated REUS also ran for 36 ns, using structures from both previous rounds and newly generated seeds. The presented data for the WT systems are compiled from the second to the last round of REUS. The total aggregation times for the REUS simulations were 65, 91, and 12 μs for the WT glycosylated, un-glycosylated, and diproline mutated systems, respectively.

4.3 Results

4.3.1 Glycans Influence the Energetics and Pathway of Spike Opening

We investigated the free-energy changes associated with the transition from the down state to the up state of the spike protein using REUS simulations for both fully glycosylated and un-glycosylated forms. The WT up state was modeled after the diproline mutant structure reported by Walls et al. [10] (PDB: 6VYB), and the down state was based on a structure from Cai et al. [12] (PDB: 6XR8), which lacks diproline mutations. Glycans at sites with the highest observed populations in mass spectrometry data from Crispin et al. [19] were added using the GLYCAM Web server (https://glycam.org) [39, 40]. Illustrations of these models are provided in Fig. 4.1a,b, with detailed descriptions in the Methods section.

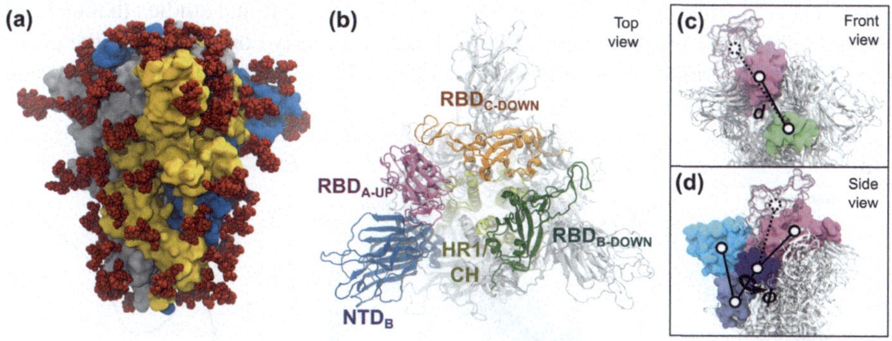

Fig. 4.1 S-protein of SARS-CoV-2. (**a**) The trimeric S-protein in the all-down state, colored by protomer. Glycans are shown as red spheres. (**b**) Top view of the S-protein in the one-up state. Important domains of the spike are highlighted, including the N-terminal domain (NTD, 14–306), the receptor binding domains (RBD, 336–518), the heptad repeat 1 (HR1, 908–986), and the central helix (CH, 987–1035). (**c**, **d**) The two collective variables defined to describe the opening of RBD-A include: (**c**) the center-of-mass distance d between RBD-A (336–518, pink) and SD1-B (531–592, lime), and (**d**) the dihedral angle ϕ formed by the center of mass of the domains RBD-A (336–518, pink), SD1-A (531–592, purple), SD2-A (593–677, ice blue), and NTD-A (27–307, cyan). RBD-A in both the down (solid pink) and up (transparent pink) states are shown. Reproduced with permission from Pang, Y. T.; Acharya, A.; Lynch, D. L.; Pavlova, A.; Gumbart, J. C. "SARS-CoV-2 spike opening dynamics and energetics reveal the individual roles of glycans and their collective impact." *Commun Biol*, vol. 5, no. 1, pp. 1170, 2022

Initially, metadynamics simulations were performed on the cryo-EM structures in both the down and up states to explore the conformational space and facilitate transitions between the two states. The conformational space was characterized by two collective variables: (1) the center-of-mass distance (d) between the opening RBD-A and a stationary part of the spike designated as the pivot, specifically the neighboring subdomain 2 on chain B (SD2-B), and (2) the dihedral angle (ϕ) formed by the opening RBD-A with other stationary domains on the same protomer, including SD2-A, SD1-A, and NTD-A (Fig. 4.1c,d). The stationary domains exhibited minimal dynamics compared to the range of motion observed in RBD-A (Table C.1). Snapshots from these metadynamics simulations served as the starting points for subsequent REUS simulations, which proceeded along the same two collective variables. For the glycosylated system, simulated annealing was applied to the glycans to randomize their conformations across each simulation window. A series of REUS simulations then systematically explored different regions of the d-ϕ space, utilizing up to 1049 and 1211 windows for the glycosylated and un-glycosylated systems, respectively (Table C.2). The total simulation time amounted to 65 μs for the glycosylated system and 91 μs for the un-glycosylated system.

For the glycosylated system, the conformations corresponding to the two cryo-EM structures were identified as energy minima on the 2D PMF, as expected (Fig. 4.2a). Among these, the up state was found to be energetically less favorable than the down state, with a difference of 5.2 ± 0.1 kcal/mol (Figs. 4.2c and C.1a). This finding is consistent with several computational studies that utilized various enhanced sampling methods, which have similarly concluded that the down state is more thermodynamically stable [33, 34, 52, 53]. The distinction between

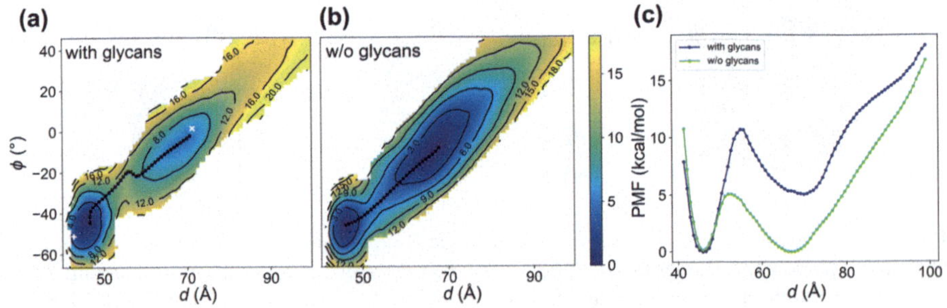

Fig. 4.2 PMFs describing the opening of RBD-A. (**a**, **b**) The 2D PMFs of the (**a**) glycosylated and (**b**) un-glycosylated systems along two collective variables, d and ϕ, defined to describe the opening of RBD-A (Fig. 4.1c,d). The location of the down- (6XR8) and up-state (6VYB) cryo-EM structures are indicated in (**a**) with a "+" and "x" sign, respectively. The black dotted line shows the MEP for each system. (**c**) The free energies are projected onto (**d**) and plotted as 1D PMFs. See also Supplementary Data 1. Reproduced with permission from Pang, Y. T.; Acharya, A.; Lynch, D. L.; Pavlova, A.; Gumbart, J. C. "SARS-CoV-2 spike opening dynamics and energetics reveal the individual roles of glycans and their collective impact." *Commun Biol*, vol. 5, no. 1, pp. 1170, 2022

the two energy wells extends beyond their depth to their breadth as well. When measuring the width of the energy wells at a level of 2.5 kcal/mol above their respective minima, the down-state well was found to range from $d = 43.1$–48.9 Å, whereas the up-state well extended from $d = 60.1$–75.6 Å. This makes the up-state well ~2.7 times wider than its counterpart. Studies by Zimmerman et al. [33] and Sztain et al. [32] also observed high flexibility in the RBD in the up state, allowing for a wide array of possible structures that exceed the dimensions typically seen in cryo-EM structures. Our PMF calculations now provide a quantitative basis for estimating the likelihood of encountering such extensively open conformations. We also computed the minimum energy path (MEP) connecting the down and up states (Fig. 4.2a) [54, 55]. The pathway predominantly follows a diagonal trajectory on the 2D PMF, with notable deviations such as a sharp increase in ϕ when exiting the down-state energy well and a slight decrease in ϕ upon entering the up-state energy well. The energy barrier between these stable conformations peaks at 11.0 ± 0.1 kcal/mol, located at $d = 55.6$ Å.

The PMF for the un-glycosylated system exhibits qualitative similarities but significant quantitative differences compared to its glycosylated counterpart (Fig. 4.2b). Notably, in the absence of glycans, the energy differential between the down and up states is negligible and within statistical error (Figs. 4.2c and C.1b). However, the width ratio along d between the energy wells of the down- and up-state remains consistent at 2.7, with the down-state well extending from $d = 43.6$–49.1 Å and the up-state well ranging from $d = 59.0$–74.1 Å. The broader energy well in the up state leads to a dominant equilibrium population of 88%, compared to 12% in the down state, despite the minimal energy difference. Additionally, the positions of the up-state energy minimum and the energy barrier are observed to shift closer to the down state, moving from $d = 70.6$–67.6 Å and from $d = 55.6$–52.6 Å, respectively, while the height of the energy barrier is reduced to 5.1 ± 0.1 kcal/mol. These findings imply that the removal of glycans not only alters the down- to up-state population ratio but also impacts the extent of RBD opening at the up-state minimum, potentially affecting the functionality of the S-protein. Previous experimental studies suggest that removal of glycans around the RBD (N234, N165, and N343) [26, 32] or reduction of glycan complexity [56] diminishes ACE2 binding.

4.3.2 Kinetics of Opening and Binding Are Altered by the Removal of Glycans

To quantify the kinetics of S-protein opening, we computed the mean first passage time (MFPT) along the MEP for the glycosylated system using the Smoluchowski diffusion equation [57, 58]. This equation models the motion of a Brownian "particle," relying on the diffusion coefficient and the free energy profile along the MEP. Additional constrained simulations were employed to determine the diffusion coefficient using the velocity autocorrelation function (VACF) [59, 60]. For the glycosylated system, the MFPT from the down state to the up state is calculated to be 1478.5 ms, while the reverse transition takes merely 0.9 ms. This contrasts with experimental observations from single-molecule FRET studies [24], which suggest that the down-to-up MFPT is overestimated by our model by $\sim 5\times$, and the reverse transition is underestimated by $\sim 100\times$. This discrepancy may arise from not fully accounting for the time required for glycans to equilibrate when calculating the diffusion coefficient. We further evaluated the MFPT for the un-glycosylated system. The observed reduction in the energy barrier within the PMF (Fig. 4.2c) correspondingly decreased the MFPT values, recorded at 143 μs for the down-to-up transition and 497 μs for the up-to-down transition.

Next, we considered the chemical dynamics according to the following paired reactions:

$$\mathrm{D + ACE2} \underset{k_{\mathrm{close}}}{\overset{k_{\mathrm{open}}}{\rightleftharpoons}} \mathrm{U + ACE2} \underset{k_{\mathrm{off}}}{\overset{k_{\mathrm{on}}}{\rightleftharpoons}} \mathrm{U{:}ACE2}$$

Here, D represents the down state of the S-protein, U represents the up state, and B denotes the bound state with ACE2 (Fig. 4.3). The opening and closing rates ($k_{\mathrm{open}}/k_{\mathrm{close}}$) were derived from the MFPTs, while the binding and unbinding rates ($k_{\mathrm{on}}/k_{\mathrm{off}}$) for RBD binding to ACE2 were sourced from Lan et al. [22]. Solving the Master equation for these reactions, with the concentration of ACE2 ($[\mathrm{ACE2}]_i$) set

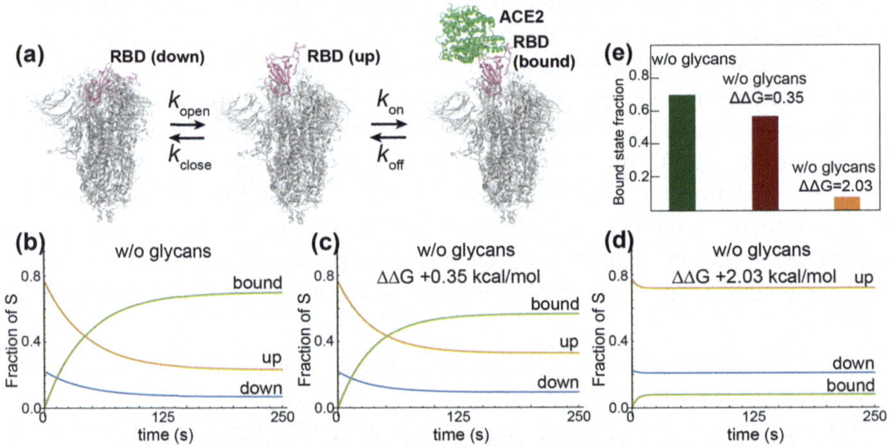

Fig. 4.3 Kinetics of S-protein opening and closing. (**a**) Transitions between RBD-down, up, and bound states are shown with their associated rates. (**b–d**) Fraction of S proteins in each state (up, down, and ACE2-bound) under different conditions, namely (**b**) with no glycans, (**c**) with no glycans and an assumed increase in the free energy of binding of RBD to ACE2 of 0.35 kcal/mol, and (**d**) with no glycans and an increase in binding free energy of 2.03 kcal/mol. (**e**) Bound-state fraction at equilibrium for the three conditions in (**b–d**). Reproduced with permission from Pang, Y. T.; Acharya, A.; Lynch, D. L.; Pavlova, A.; Gumbart, J. C. "SARS-CoV-2 spike opening dynamics and energetics reveal the individual roles of glycans and their collective impact." *Commun Biol*, vol. 5, no. 1, pp. 1170, 2022

at 15 nM (see Supplementary Information), indicated that for the un-glycosylated S-protein, the populations are 70% bound, 23% up-but-unbound, and only 7% in the down state (Fig. 4.3b). Deep mutational scanning of the RBD has shown that any mutation removing the glycan at N343 is deleterious, with inferred $\Delta\Delta G$ values ranging from +0.35 to 2.03 kcal/mol [61]. Using the open/close rates for the un-glycosylated S-protein, adjusted for the binding/unbinding rates based on this mutation data, we found that the bound-state population significantly decreases to between 8% ($\Delta\Delta G = 2.03$ kcal/mol) and 57% ($\Delta\Delta G = 0.35$ kcal/mol). Therefore, while the removal of glycans increases the favorability of the up state, the corresponding reduction in binding affinity could counteract the anticipated increase in the ACE2-bound population.

4.3.3 Glycans Interfere with Hydrogen Bonds that Stabilize the Up-state RBD

To elucidate the impact of glycans on the energetics of spike protein opening, we analyzed the interactions between the RBD and the remainder of the protein during the REUS simulations. We quantified the number of hydrogen bonds between the opening RBD-A and other protein domains, plotting these against the path parameter (Figs. 4.4a,b and C.2). Notably, the number of hydrogen bonds generally decreased as RBD-A transitioned from closed to open, with an increase in interactions observed after crossing the energy barrier—this effect was more pronounced in the glycosylated system. In the glycosylated system, RBD-A consistently formed more hydrogen bonds, especially in the down state. However, when only considering protein-protein interactions, the un-glycosylated RBD-A established more connections with other protein domains. This decline in hydrogen bonds is primarily attributed to RBD-A distancing itself from other parts of the trimeric S-protein, such as adjacent RBDs and S2 domains, as it opens. Yet, this spatial separation alone does not completely account for the observed increase in interactions post-energy barrier or the distinct differences noted between glycosylated and un-glycosylated simulations.

4.3.4 Analysis of Hydrogen-Bond Changes Along the MEP

A detailed analysis of the changes in hydrogen-bond composition along the MEP sheds light on the role of glycans in the transition from the down to up state of the spike protein. In the down state, RBD-A predominantly forms hydrogen bonds with the other two RBDs and the HR1/CH helix within the S2 unit. As the protein transitions to the up state, these interactions shift markedly to involve

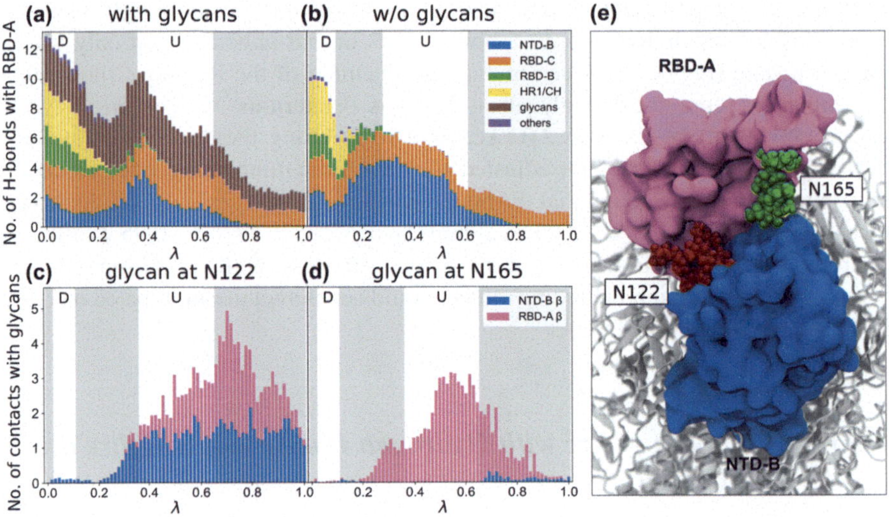

Fig. 4.4 Hydrogen bond and contact analysis reveals the different interaction patterns with and without glycans. (**a**, **b**) Along the MEPs as indicated by the path parameter λ (see Fig. C.2), the number of hydrogen bonds formed between the opening RBD-A and the rest of the spike are counted and classified by domain for the (**a**) glycosylated and the (**b**) un-glycosylated systems. The locations of the down- and up-state energy wells are shown with a white background while other regions are shaded in grey. (**c**, **d**) The average number of contacts for the glycosylated system formed between the glycans at (**c**) N122 and (**d**) N165 and the neighboring β-strand from NTD-B (165–172) and RBD-A (353–360), which would otherwise form hydrogen bonds with each other if not separated by the glycans. (**e**) Snapshot from the REUS simulation showing the glycans at N165 and N122 disrupting hydrogen bond formation between RBD-A and NTD-B, destabilizing the up state compared to the un-glycosylated system. Reproduced with permission from Pang, Y. T.; Acharya, A.; Lynch, D. L.; Pavlova, A.; Gumbart, J. C. "SARS-CoV-2 spike opening dynamics and energetics reveal the individual roles of glycans and their collective impact." *Commun Biol*, vol. 5, no. 1, pp. 1170, 2022

the neighboring NTD-B and RBD-C, a change that contributes to the stabilization dynamics observed in this conformation (Figs. 4.4a,b and C.3).

The introduction of glycans leads to additional interactions that stabilize the protein; however, these glycans also compete with protein-protein hydrogen bonds. In the compact, down-state structure, protein-protein hydrogen bonds are shielded by glycans, which fosters new interactions on the protein surface, ultimately resulting in an overall increase in hydrogen bonds involving RBD-A compared to the un-glycosylated system. Conversely, in the up state, this protective effect is diminished. When RBD-A is in the up and exposed configuration, its interactions with NTD-B are notably reduced. Most hydrogen bonds between RBD-A and NTD-B are replaced by interactions with glycans at N165 and N122 once RBD-A is sufficiently elevated (Fig. 4.4c–e).

Consequently, glycans indirectly support the preference for the down state by contributing to the higher free energy observed in the up state of the glycosylated system. This mechanism underscores the complex interplay between glycan shielding and protein dynamics, illustrating how glycans can influence the functional states of the S-protein.

4.3.5 Extended Analysis of Hydrogen Bonds Across the 2D Collective Variable Space

Extending our analysis of hydrogen bonds beyond the MEP to the entire 2D collective variable space has deepened our understanding of the differences in free energy landscapes between the glycosylated and un-glycosylated systems from the REUS simulations. We plotted the average number of hydrogen bonds between the opening RBD-A and its neighboring NTD-B against the two collective variables, d and ϕ. This analysis indicates that the increase in interactions between RBD-A and NTD-B is not confined to states along the MEP but extends across a broad region on the $+d$-side of the energy barrier (Fig. C.4a).

Interestingly, the number of hydrogen bonds between RBD-A and NTD-B peaks at $\phi = 40°$ and gradually diminishes as ϕ increases. This trend supports the observation that as RBD-A moves further from NTD-B, glycans start to interfere, blocking interactions between the two domains. Conversely, on the $-d$-side of the energy barrier, the scenario is reversed for hydrogen bonds between RBD-A and RBD-C, where the two domains are closest when ϕ is large, resulting in a maximal number of hydrogen bonds (Fig. C.4b). This pattern elucidates the characteristic kink in the MEP when the glycosylated system crosses the energy barrier. To optimize the stabilizing hydrogen bonds for RBD-A, the spike protein exits the down-state energy well with a large ϕ to maintain maximum contact with RBD-C, then abruptly transitions to a smaller ϕ upon entering the up-state energy well to maximize interactions with NTD-B.

This requirement for RBD-A to switch from hydrogen bonding with RBD-C to NTD-B necessitates a significant conformational change in the orientation of RBD-A, contributing to the pronounced energy barrier between the down and up states. Stabilizing interactions are present on both sides of the barrier—with RBD-C on the $-d$ side and with NTD-B on the $+d$ side. The absence of these stabilizing interactions at $d = 54.9$ Å is what forms the energy barrier. This challenge is navigated by crossing the barrier at a high ϕ, where RBD-A maintains maximum hydrogen bonds with RBD-C, resulting in a distinct MEP in the potential of mean force (PMF) for the glycosylated system, unlike the relatively linear MEP observed in the un-glycosylated system.

4.3.6 Distinct Glycan Contacts with RBD-A Stabilize Both the Up and Down States

To probe the long-term dynamics of glycans, we performed two 2-μs equilibrium simulations for both glycosylated and un-glycosylated systems. When mapped onto the two collective variables utilized in REUS simulations, all three protomers in the down state of both replicas maintained their conformations throughout, regardless of glycan presence (Fig. C.5). Notably, no transitions between the up and down states were observed during these simulations, underscoring the stability of the down state.

However, the up states demonstrated significantly greater flexibility compared to the down states in both glycosylated and un-glycosylated forms. This observation mirrors the disparity in the size of the energy wells identified in the REUS simulations. In the un-glycosylated system, RBD-A navigated around the up-state energy minimum without showing a clear directional preference. In contrast, in the glycosylated system, RBD-A exhibited a tendency to migrate towards the $-d$ direction from the up-state minimum along the MEP. This behavior corresponds with findings from REUS simulations, which suggest that the up-state energy well for the glycosylated system has a gentler gradient towards the energy barrier than away from it. Meanwhile, the energy well of the un-glycosylated system appears more symmetric (Fig. 4.2c).

These results indicate that glycans not only enhance the structural integrity of specific states but also influence the directional tendencies of RBD-A movement within the energy landscape, particularly in the up state. This nuanced role of glycans contributes to the stability and dynamism observed in the spike protein's functional states.

4.3.7 Interactions of Glycans Surrounding RBD-A in Stabilizing the Spike Protein

In our investigation of the interactions of glycans surrounding RBD-A, specifically at N165, N234, and N343 of the neighboring chain B, we utilized data from equilibrium simulations and the REUS trajectories along the MEP. This analysis complements recent studies by Amaro and colleagues, who utilized all-atom MD simulations and mutational experiments to explore the roles of these glycans in facilitating spike protein opening [26, 32]. Consistent with their findings, we observed the inward swing of the glycan at N234 during spike opening and the glycan gating effect facilitated by the glycan at N343.

Our simulations further demonstrated that glycans at N165 and N343 contribute to the stabilization of both the up and down states of RBD-A. Specifically, in the down state, glycans at N165 and N343 envelop the RBM, maintaining its position (Fig. 4.5a). This interaction mirrors observations from another study where an artificially introduced glycan at N370 encircled the down-state RBD, functioning

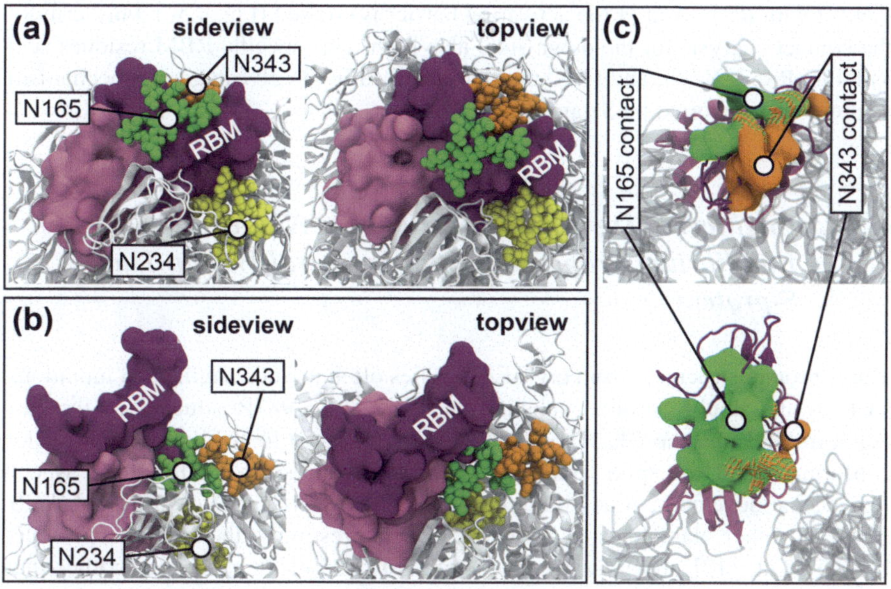

Fig. 4.5 Glycan-protein interactions. Representative locations of S-protein glycans at N165, N234, and N343 in S-protein (**a**) down and (**b**) up states extracted from the REUS trajectories. (**c**) Surface representation of RBD residues in (above) down and (below) up states that make contact with glycans at N165 (green) and N343 (orange). The cross-hatch pattern indicates contacts with both glycans. The up and down states of the S-protein were selected from the MEP. Reproduced with permission from Pang, Y. T.; Acharya, A.; Lynch, D. L.; Pavlova, A.; Gumbart, J. C. "SARS-CoV-2 spike opening dynamics and energetics reveal the individual roles of glycans and their collective impact." *Commun Biol*, vol. 5, no. 1, pp. 1170, 2022

like a "shoelace" to secure its position [21]. Remarkably, the glycan at N165 also supports the stabilization of the up-state RBD, reinforcing part of the exposed RBM (Fig. 4.5b). The inter-glycan contact between N343 and N165 is significantly more pronounced in the up state than in the down state (Fig. C.6), suggesting that the interaction between these glycans indirectly stabilizes the open state of RBD-A (Fig. 4.5b).

These glycans play a crucial role in preventing the flexible RBM from binding to the neighboring down-state RBD, thereby keeping it accessible to ACE2. This protective mechanism was notably absent in one of our equilibrium simulations of the un-glycosylated system, where the RBM became inadvertently attached to an adjacent down-state RBD and rendered inaccessible to ACE2 (Fig. C.7).

The observations described are corroborated by contact analysis of the REUS trajectories along the MEP. The glycans at N165 and N343 both form interactions with the RBM in the down state, evidenced by the average contact values between RBD-A and the glycans at N165 and N343 (Figs. C.8 and C.9). As the spike transitions to the open state, the glycan at N165 alters its interaction to connect with residues below the RBM of RBD-A, while the glycan at N343 shows reduced contact with RBD-A after the activation barrier is crossed (Fig. 4.5c). Furthermore, the contact analysis for the down state also highlights specific RBM residues such as F456, R457, Y489, and F490 that are involved in the glycan gating mechanism, as previously reported by Sztain et al. [32].

4.3.8 Engineered Diproline Mutations Do Not Affect the Equilibrium Properties of the Ancestral Wuhan-Hu-1 S-protein

The S-protein structures from cryo-EM studies often include engineered mutations, such as the double proline substitution at K986P/V987P, situated in the turn between HR1 and the CH. These mutations are known to stabilize the pre-fusion conformation of the spike and enhance protein expression, both crucial aspects for vaccine development as highlighted by prior research on SARS-CoV, MERS-CoV, and other class I fusion proteins [62]. Interestingly, the structure provided by Cai et al. [12] preserves the ancestral WT K986/V987 sequence rather than the commonly used diproline mutations, revealing a more compact arrangement of the S1 subunits compared to earlier solubilized ectodomain structures [12]. This configuration change is attributed to the loss of a putative salt bridge between K986 and an aspartic acid (D427/D428) on an adjacent protomer, which has been linked to less tightly packed structures observed in mutant constructs [12].

Furthermore, Gobeil et al. [11] observed that, in the D614 S-protein variant with the furin cleavage site removed, the structural characteristics, ACE2 binding, thermal stability, and antibody binding of the spike with prolines are remarkably akin to those with the WT K986/V987 pair. In our equilibrium simulations, the aforementioned salt bridge exhibited an average occupancy of about 30% across all three protomers in the two independent trajectories. This reduced occupancy stems from competitive intra-protomer salt bridges involving K986 and nearby acidic residues—E748, E990, and D985—with occupancies in the range of approximately 15–25%, suggesting transient interactions between K986 and the RBD aspartic acids (D427/D428). These competing interactions, which potentially modulate the structural integrity and functional dynamics of the S-protein, are detailed in Fig. C.10.

We expanded our analysis to include REUS simulations with the diproline mutation in the un-glycosylated system. The resultant PMF shows only a modest energy difference between the down and up RBD states, with a slight favoring of the up state (Fig. C.11), mirroring the pattern observed in the WT PMF (Fig. 4.2b) for the ancestral D614 constructs. This similarity between the energy wells of the down and up states in the presence of the diproline mutation suggests that these modifications induce only a modest perturbation in their relative populations within these un-glycosylated systems. Unlike the minimal changes in the minima, the energy barrier in the diproline mutant appears elevated, hinting at altered opening kinetics.

Given the transient nature of the K986 salt bridges and the comparable relative energies of the down and up RBD states, another significant impact of the proline mutations is likely their propensity to disrupt and distort α-helices. The HR1-CH turn region, which undergoes substantial conformational changes during the transition from pre- to post-fusion states, typically forms an extended α-helix [12]. However, the introduction of proline residues, which are known to disrupt or kink α-helices, impedes the formation of the long α-helix characteristic of the post-fusion conformation [62]. Complementary simulations by Wang et al. [63] that explore the transition from the pre- to post-fusion states have demonstrated that while the WT sequence at positions 986/987 tends to form helical structures, replacement with prolines disrupts this secondary structure development.

4.3.9 Differential Impact of Glycans on Epitope Exposure

The role of glycans in modulating epitope exposure on the SARS-CoV-2 S-protein is significant, especially given the large number of neutralizing antibodies targeting this virus. According to recent studies, nearly ten thousand neutralizing antibodies have been identified that interact with the S-protein RBD [64, 65]. The continuously updated CoV-AbDab database reports that out of 9906 neutralizing antibodies targeting the S-protein, 3955 are directed at non-RBD epitopes [66]. This disparity highlights the immunological importance of the RBD as the most prominent target, although significant attention is also given to the NTD and other non-RBD epitopes [67].

One key advantage of targeting non-RBD epitopes is their accessibility even when the RBD is in the down conformation [20, 68]. In contrast, the RBD is typically shielded by extensive glycan coverage when in the down conformation, making it recognizable only in its up conformation [26]. This glycan shielding significantly influences antibody recognition, suggesting a strategic advantage in targeting epitopes that remain exposed regardless of the RBD conformation. This differential impact of glycans on epitope exposure not only affects the efficacy of neutralizing antibodies but also informs vaccine design, where epitope accessibility can dictate immune response effectiveness.

4.3.10 Impact of Spike-Opening Dynamics on Epitope Exposure

To elucidate the impact of spike-opening dynamics on the accessibility of various epitopes, we quantitatively assessed the accessible surface area (ASA) of several epitopes along the MEP using data from REUS simulations. We selected seven different antibodies known to target epitopes across RBD-A and NTD-B of the S-protein, including regions harboring cryptic epitopes [16, 17, 69–73]. Detailed information about these antibodies is available in the Supplementary Information.

We conducted antibody accessible surface area (AbASA) calculations both including and excluding glycans, utilizing a consistent sampling approach with a 7-Å probe size. This probe size was selected based on its efficacy in previous studies, where it has been shown to be suitable for revealing moderately exposed crevices and is particularly effective for cryptic epitopes [20, 33]. While this moderate probe size can sometimes overestimate exposure due to its ability to penetrate smaller gaps [20], it provides a balanced approach to accurately gauge the accessibility of epitopes that are crucial for neutralizing antibody binding.

The AbASA results either remain stable or increase significantly when RBD-A transitions to the up state. Notably, the epitope of the STE90-C11 antibody shows a considerable increase in AbASA during the closed-to-open transition [72]. Conversely, the CR3022 antibody, which binds to a cryptic epitope, shows that this region is completely covered in the down state and only slightly exposed in the up state [70]. The protein residues consistently cover this epitope across the spike opening path for MEPs with and without glycans (Fig. C.12).

Remarkably, the AbASA for the 2–4 antibody does not change during the down-to-up transition of the RBD, despite the epitope being located in the RBM (Fig. 4.6a). This observation suggests that the entire RBM does not become fully exposed even in the up state. Similarly, the AbASA for the epitope of the 4A8 antibody, located in the N-terminus, does not change significantly during spike opening, as its location does not undergo substantial conformational changes [17]. For the S2M11 antibody, the AbASA remains unchanged along the spike opening path; its epitope, partially on the RBD of the adjacent protomer, does not open in

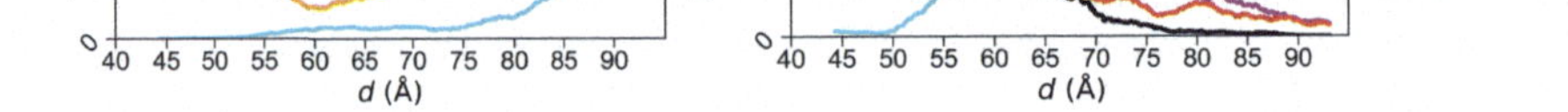

Fig. 4.6 Epitope analysis for selected antibodies. (**a**) Exposed area on antibody epitopes (AbASA) in the presence of protein and glycans. (**b**) Surface area of epitopes covered by glycans along the MEP quantified by subtracting the two AbASA values calculated with and without glycans. All accessible surface area calculations were performed using a 7-Å probe. Reproduced with permission from Pang, Y. T.; Acharya, A.; Lynch, D. L.; Pavlova, A.; Gumbart, J. C. "SARS-CoV-2 spike opening dynamics and energetics reveal the individual roles of glycans and their collective impact." *Commun Biol*, vol. 5, no. 1, pp. 1170, 2022

our simulations. The S2M11 antibody's epitope slightly overlaps with the RBD of one protomer (residues 440, 441, and 444), while the majority aligns with the binding site of the ACE2 glycan at N322 on the RBD (residues 369, 371 to 374, and 440) [74]. Interestingly, the epitope of the S309 antibody becomes less exposed in the open state without an increase in glycan coverage (Fig. 4.6b), indicating that parts of the RBD become less accessible in the up state compared to the down state [71].

These findings highlight that the primary regions of variant mutations in the NTD and RBD significantly affect the effectiveness of neutralizing antibodies, such as STE90-C11, 4A8, S2M11, and BD-368-2, particularly against Omicron BA.1 [75–77] and increased S309 escape by Omicron BA.2, BA.4/BA.5 [78]. The proximity of these mutations to the N343 glycan underscores the critical role glycans play in immune response modulation.

4.4 Discussion

SARS-CoV-2 infection initiates when the ACE2 receptor binds to the S-protein, which dynamically interconverts between up and down conformations; only the up state permits ACE2 binding. Understanding this conformational interconversion is crucial for controlling infection initiation. To this end, we computed a two-dimensional energy landscape of this up-down interconversion using extensive REUS simulations, aggregating a total of 160 μs of computational analysis. Our results underscore the significant role of S-protein glycans in influencing opening dynamics. Notably, the free-energy barrier for spike opening is approximately 7 kcal/mol higher in glycosylated systems, demonstrating a substantial glycan impact that favors the down state. Conversely, without glycans, the energy difference between states diminishes to below statistical significance, indicating a near-balanced population between the conformations.

The dynamics of the spike protein are central to its functionality. While cryo-EM structures provide snapshots of these states, they do not capture the transitions or the intermediate states effectively. MD simulations serve as a crucial complement to these experimental approaches, providing insights into the kinetic effects of glycan modifications and the dynamic balance of the spike's conformations. For example, recent cryo-EM studies of the Omicron variant's ectodomain predominantly show a 1-up RBD configuration, suggesting a shift in the equilibrium towards more open states under certain experimental conditions or viral constructs [79–81].

Our analysis extends to the exposure of neutralizing epitopes along the spike-opening path. Antibodies targeting the RBD and NTD regions, crucial for neutralization, show varying degrees of exposure due to glycan shielding and protein conformational changes. For instance, the epitope targeted by the STE90-C11 antibody becomes significantly more accessible in the up state, while others like the S309 epitope remain exposed in both states due to their locations away from the RBM. The glycan at N343, termed the "glycan gate," plays a pivotal role by stabilizing the RBD in both the down and up states, affecting the exposure and accessibility of these critical neutralizing sites [32].

Furthermore, the ongoing evolution of the spike under immunological pressure, exemplified by variants such as Omicron, underscores the necessity of continually assessing the spike's structural and dynamic profiles. These variants exhibit increased transmissibility and potential for immune escape, prompting ongoing adjustments in vaccine and therapeutic strategies. The detailed energetics and kinetics of spike opening, as revealed by our computational analyses, are instrumental in designing effective countermeasures against evolving SARS-CoV-2 challenges.

In summary, our study provides a comprehensive atomic-level description of the spike-opening dynamics, highlighting the critical influence of glycans and the implications for ACE2 binding and antibody recognition. These insights are crucial for the ongoing development of targeted therapies and vaccines in the face of the persistent COVID-19 pandemic.

References

1. B. Hu, H. Guo, P. Zhou, Z.-L. Shi, Characteristics of SARS-CoV-2 and COVID-19. Nat. Rev. Microbiol. **19**, 141–154 (2020)
2. Y. Liu, K. Wang, T.F. Massoud, R. Paulmurugan, SARS-CoV-2 vaccine development: an overview and perspectives. ACS Pharmacol. Transl. Sci. **3**(5), 844–858 (2020)
3. P.C. Taylor, A.C. Adams, M.M. Hufford, I. de la Torre, K. Winthrop, R.L. Gottlieb, Neutralizing monoclonal antibodies for treatment of COVID-19. Nat. Rev. Immunol. **21**(6), 382–393 (2021)
4. D.R. Owen et al., An oral SARS-CoV-2 M^{pro} inhibitor clinical candidate for the treatment of COVID-19. Science **374**, 1586–1593 (2021)
5. J.E. Bowen et al., Omicron spike function and neutralizing activity elicited by a comprehensive panel of vaccines. Science **377**(6608), 890–894 (2022)
6. D.K. Bonilla-Aldana et al., Bats in ecosystems and their Wide spectrum of viral infectious potential threats: SARS-CoV-2 and other emerging viruses. Int. J. Infect. Dis. **102**, 87–96 (2021)
7. M. Hoffmann et al., SARS-CoV-2 cell entry depends on ACE2 and TMPRSS2 and is blocked by a clinically proven protease inhibitor. Cell **181**(2), 271–280 (2020)
8. Q. Wang et al., Structural and functional basis of SARS-CoV-2 entry by using human ACE2. Cell **181**(4), 894–904 (2020)
9. D. Wrapp et al., Cryo-EM structure of the 2019-nCoV spike in the prefusion conformation. Science **367**(6483), 1260–1263 (2020)
10. A.C. Walls, Y.J. Park, M.A. Tortorici, A. Wall, A.T. McGuire, D. Veesler, Structure, function, and antigenicity of the SARS-CoV-2 spike glycoprotein. Cell **181**(2), 281–292 (2020)
11. S.M. Gobeil et al., D614G mutation alters SARS-CoV-2 Spike conformation and enhances protease cleavage at the S1/S2 junction. Cell Rep. **34**(2), 108630 (2021)
12. Y. Cai et al., Distinct conformational states of SARS-CoV-2 spike protein. Science **369**(6511), 1586–1592 (2020)
13. Z. Ke et al., Structures and distributions of SARS-CoV-2 spike proteins on intact virions. Nature **588**, 498–502 (2020)
14. W.T. Harvey et al., SARS-CoV-2 variants, spike mutations and immune escape. Nat. Rev. Microbiol. **19**, 409–424 (2021)

15. G. Cerutti et al., Potent SARS-CoV-2 neutralizing antibodies directed against spike N-terminal domain target a single supersite. Cell Host Microbe **29**(5), 819–833 (2021)
16. L. Liu et al. Potent neutralizing antibodies against multiple epitopes on SARS-CoV-2 spike. Nature **584**, 450–456 (2020)
17. X. Chi et al., A neutralizing human antibody binds to the N-terminal domain of the Spike protein of SARS-CoV-2. Science **369**, 650–655 (2020)
18. W. Tai et al., Characterization of the receptor-binding domain (RBD) of 2019 novel coronavirus: implication for development of RBD protein as a viral attachment inhibitor and vaccine. Cell Mol. Immunol. **17**(6), 613–620 (2020)
19. Y. Watanabe et al., Vulnerabilities in coronavirus glycan shields despite extensive glycosylation. Nat. Commun. **11**(1), 2688 (2020)
20. O.C. Grant, D. Montgomery, K. Ito, R.J. Woods, Analysis of the SARS-CoV-2 spike protein glycan shield reveals implications for immune recognition. Sci. Rep. **10**(1), 14991 (2020)
21. A.M. Harbison, C.A. Fogarty, T.K. Phung, A. Satheesan, B.L. Schulz, E. Fadda, Fine-tuning the spike: role of the nature and topology of the glycan shield in the structure and dynamics of the SARS-CoV-2 S. Chem. Sci. **13**(2), 386–395 (2022)
22. J. Lan et al., Structure of the SARS-CoV-2 spike receptor-binding domain bound to the ACE2 receptor. Nature **581**, 215–220 (2020)
23. I. Berger and C. Schaffitzel, The SARS-CoV-2 spike protein: balancing stability and infectivity. Cell Res. **30**(12), 1059–1060 (2020)
24. M. Lu et al., Real-time conformational dynamics of SARS-CoV-2 spikes on virus particles. Cell Host Microbe **28**(6), 880–891 (2020)
25. Z. Yang et al., SARS-CoV-2 variants increase kinetic stability of open spike conformations as an evolutionary strategy. mBio **13**, e0322721 (2022)
26. L. Casalino et al. Beyond shielding: the roles of glycans in the SARS-CoV-2 spike protein. ACS Cent. Sci. **6**(10), 1722–1734 (2020)
27. D. Ray, L. Le, I. Andricioaei, Distant residues modulate conformational opening in SARS-CoV-2 spike protein. Proc. Natl. Acad. Sci. U. S. A. **118**, e2100943118 (2021)
28. Y.K. Choi et al., Structure, dynamics, receptor binding, and antibody binding of the fully glycosylated full-length SARS-CoV-2 spike protein in a viral membrane. J. Chem. Theory Comput. **17**(4), 2479–2487 (2021)
29. T. Mori, J. Jung, C. Kobayashi, H.M. Dokainish, S. Re, Y. Sugita, Elucidation of interactions regulating conformational stability and dynamics of SARS-CoV-2 S-protein. Biophys. J. **120**(6), 1060–1071 (2021)
30. M. Gur, E. Taka, S.Z. Yilmaz, C. Kilinc, U. Aktas, M. Golcuk, Conformational transition of SARS-CoV-2 spike glycoprotein between its closed and open states. J. Chem. Phys. **153**(7), 075101 (2020)
31. V. Govind Kumar, D.S. Ogden, U.H. Isu, A. Polasa, J. Losey, M. Moradi, Prefusion spike protein conformational changes are slower in SARS-CoV-2 than in SARS-CoV-1. J. Biol. Chem. **298**(4), 101814 (2022)
32. T. Sztain et al. A glycan gate controls opening of the SARS-CoV-2 spike protein. Nat. Chem. **13**(10), 963–968 (2021)
33. M.I. Zimmerman et al., SARS-CoV-2 simulations go exascale to predict dramatic spike opening and cryptic pockets across the proteome. Nat. Chem. **13**, 651–659 (2021)
34. L. Fallon et al., Free energy landscapes from SARS-CoV-2 spike glycoprotein simulations suggest that RBD opening can be modulated via interactions in an allosteric pocket. J. Am. Chem. Soc. **143**, 11349–11360 (2021)
35. Z.F. Brotzakis, T. Löhr, M. Vendruscolo, Determination of intermediate state structures in the opening pathway of SARS-CoV-2 spike using cryo-electron microscopy. Chem. Sci. **12**(26), 9168–9175 (2021)
36. G. Mashayekhi, J. Vant, A. Polavarapu, A. Ourmazd, A. Singharoy, Energy landscape of the SARS-CoV-2 reveals extensive conformational heterogeneity. Curr. Res. Struct. Biol. **4**, 68–77 (2022)

37. W. Humphrey, A. Dalke, K. Schulten, VMD: visual molecular dynamics. J. Mol. Graph. **14**(1), 33–38 (1996)
38. A. Waterhouse et al., SWISS-MODEL: homology modelling of protein structures and complexes. Nucleic Acids Res. **46**, W296–W303 (2018)
39. D.F. Thieker, J.A. Hadden, K. Schulten, R.J. Woods, 3D implementation of the symbol nomenclature for graphical representation of glycans. Glycobiology **26**(8), 786–787 (2016)
40. A. Varki et al., Symbol nomenclature for graphical representations of glycans. Glycobiology **25**, 1323–1324 (2015)
41. J.C. Phillips et al., Scalable molecular dynamics with NAMD. J. Comput. Chem. **26**(16), 1781–1802 (2005)
42. J.C. Phillips et al., Scalable molecular dynamics on CPU and GPU architectures with NAMD. J. Chem. Phys. **153**, 044 130 (2020)
43. J. Huang et al., CHARMM36m: an improved force field for folded and intrinsically disordered proteins. Nat. Methods **14**(1), 71–73 (2017)
44. O. Guvench, E.R. Hatcher, R.M. Venable, R.W. Pastor, A.D. Mackerell, CHARMM additive all-atom force field for glycosidic linkages between hexopyranoses. J. Chem. Theory Comput. **5**(9), 2353–2370 (2009)
45. W.L. Jorgensen, J. Chandrasekhar, J.D. Madura, R.W. Impey, M.L. Klein, Comparison of simple potential functions for simulating liquid water. J. Chem. Phys. **79**(2), 926–935 (1983)
46. S.E. Feller, Y. Zhang, R.W. Pastor, B.R. Brooks, Constant pressure molecular dynamics simulation: the Langevin piston method. J. Chem. Phys. **103**(11), 4613–4621 (1995)
47. C.W. Hopkins, S. Le Grand, R.C. Walker, A.E. Roitberg, Long-time-step molecular dynamics through hydrogen mass repartitioning. J. Chem. Theory Comput. **11**(4), 1864–1874 (2015)
48. C. Balusek et al., Accelerating membrane simulations with hydrogen mass repartitioning. J. Chem. Theory Comput. **15**(8), 4673–4686 (2019)
49. T. Darden, D. York, L. Pedersen, Particle mesh Ewald: An $N \cdot \log(N)$ method for Ewald sums in large systems. J. Chem. Phys. **98**(12), 10089–10092 (1993)
50. G. Fiorin, M.L. Klein, J. Hénin, Using collective variables to drive molecular dynamics simulations. Mol. Phys. **111**(22–23), 3345–3362 (2013)
51. M.R. Shirts, J.D. Chodera, Statistically optimal analysis of samples from multiple equilibrium states. J. Chem. Phys. **129**(12), 124105 (2008)
52. Y. Wu, R. Qian, Y. Yang, Y. Sheng, W. Li, W. Wang, Activation pathways and free energy landscapes of the SARS-CoV-2 spike protein. ACS Omega **6**(36), 23432–23441 (2021)
53. H.M. Dokainish, S. Re, T. Mori, C. Kobayashi, J. Jung, Y. Sugita, The inherent flexibility of receptor binding domains in SARS-CoV-2 spike protein. Elife **11**, e75720 (2022)
54. B. Ensing, A. Laio, M. Parrinello, M.L. Klein, A recipe for the computation of the free energy barrier and the lowest free energy path of concerted reactions. J. Phys. Chem. B **109**(14), 6676–6687 (2005)
55. M. Moradi, V. Babin, C. Roland, T.A. Darden, C. Sagui, Conformations and free energy landscapes of polyproline peptides. Proc. Natl. Acad. Sci. U. S. A. **106**(49), 20746–20751 (2009)
56. K.M. Bouwman et al., Multimerization- and glycosylation-dependent receptor binding of SARS-CoV-2 spike proteins. PLoS. Pathog. **17**(2), e1009282 (2021)
57. M.V. Smoluchowski, Über Brownsche Molekularbewegung unter Einwirkung äußerer Kräfte und deren Zusammenhang mit der verallgemeinerten Diffusionsgleichung. Ann. Phys. **353**(24), 1103–1112 (1916)
58. A. Fakharzadeh, M. Moradi, Effective Riemannian diffusion model for conformational dynamics of biomolecular systems. J. Phys. Chem. Lett. **7**(24), 4980–4987 (2016)
59. T.B. Woolf, B. Roux, Molecular dynamics simulation of the gramicidin channel in a phospholipid bilayer. Proc. Natl. Acad. Sci. U. S. A. **91**(24), 11631–11635 (1994)
60. K. Gaalswyk, E. Awoonor-Williams, C.N. Rowley, Generalized Langevin methods for calculating transmembrane diffusivity. J. Chem. Theory Comput. **12**(11), 5609–5619 (2016)
61. T.N. Starr et al. Deep mutational scanning of SARS-CoV-2 receptor binding domain reveals constraints on folding and ACE2 binding. Cell **182**(5), 1295–1310 (2020)

62. R.W. Sanders, J.P. Moore, Virus vaccines: proteins prefer prolines. Cell Host Microbe **29**(3), 327–333 (2021)
63. Y. Wang et al., Receptor binding may directly activate the fusion machinery in coronavirus spike glycoproteins (2021). bioRxiv, https://doi.org/10.1101/2021.05.10.443496
64. D. Corti, L.A. Purcell, G. Snell, D. Veesler, Tackling COVID-19 with neutralizing monoclonal antibodies. Cell **184**, 3086–3108 (2021)
65. Y. Valdes-Balbin et al., Molecular aspects concerning the use of the SARS-CoV-2 receptor binding domain as a target for preventive vaccines. ACS Cent. Sci. **7**, 757–767 (2021)
66. M.I. Raybould, A. Kovaltsuk, C. Marks, C.M. Deane, CoV-AbDab: the coronavirus antibody database. Bioinformatics **37**(5), 734–735 (2021)
67. W.N. Voss et al. Prevalent, protective, and convergent IgG recognition of SARS-CoV-2 non-RBD spike epitopes. Science **372**, 1108–1112 (2021)
68. M. Sikora, S. von Bülow, F.E. Blanc, M. Gecht, R. Covino, G. Hummer, Computational epitope map of SARS-CoV-2 spike protein. PLoS Comput. Biol. **17**(4), e1008790 (2021)
69. S. Du et al., Structurally resolved SARS-CoV-2 antibody shows high efficacy in severely infected hamsters and provides a potent cocktail pairing strategy. Cell **183**, 1013–1023 (2020)
70. M. Yuan et al., A highly conserved cryptic epitope in the receptor binding domains of SARS-CoV-2 and SARS-CoV. Science **368**(6491), 630–633 (2020)
71. D. Pinto et al., Cross-neutralization of SARS-CoV-2 by a human monoclonal SARS-CoV antibody. Nature **583**, 290–295 (2020)
72. F. Bertoglio et al., SARS-CoV-2 neutralizing human recombinant antibodies selected from pre-pandemic healthy donors binding at RBD-ACE2 interface. Nat. Commun. **12**, 1577 (2021)
73. M.A. Tortorici et al., Ultrapotent human antibodies protect against SARS-CoV-2 challenge via multiple mechanisms. Science **370**, 950–957 (2020)
74. A. Acharya, D.L. Lynch, A. Pavlova, Y.T. Pang, J. Gumbart, ACE2 glycans preferentially interact with SARS-CoV-2 over SARS-CoV. Chem. Commun. **57**, 5949–5952 (2021)
75. S.R. Schulz et al., Augmented neutralization of SARS-CoV-2 Omicron variant by boost vaccination and monoclonal antibodies. Eur. J. Immunol. **52**(6), 970–977 (2022)
76. D. Mannar et al., SARS-CoV-2 Omicron variant: antibody evasion and cryo-EM structure of spike protein-ACE2 complex. Science **375**(6582), 760–764 (2022)
77. Y. Cao et al., Omicron escapes the majority of existing SARS-CoV-2 neutralizing antibodies. Nature **602**(7898), 657–663 (2022)
78. Y. Cao et al., BA.2.12.1, BA.4 and BA.5 escape antibodies elicited by Omicron infection. Nature **608**(7923), 593–602 (2022)
79. G. Ye, B. Liu, F. Li, Cryo-EM structure of a SARS-CoV-2 omicron spike protein ectodomain. Nat. Commun. **13**(1), 1214 (2022)
80. Q. Hong et al., Molecular basis of receptor binding and antibody neutralization of Omicron. Nature **604**(7906), 546–552 (2022)
81. J. Zhang et al. Structural and functional impact by SARS-CoV-2 Omicron spike mutations. Cell Rep. **39**(4), 110729 (2022)

Chapter 5
Conclusions and Future Work

This thesis has presented a comprehensive exploration of the conformational dynamics of three distinct biomolecules: the hepatitis B virus (HBV) capsid assembly modulator AT130, the pertactin passenger domain, and the SARS-CoV-2 spike protein. Utilizing advanced simulation techniques and enhanced sampling, we have explored the complex behaviors of these biomolecules, each representing a unique facet of biological complexity. In the process, I also updated the open source parameterization tool, Force Field Toolkit (ffTK), to allow the use of the new σ-hole particle (LP) from CHARMM General Force Field (CGenFF) 4.0. Our investigations spanned different scales and complexities, demonstrating the versatility and applicability of simulations and enhanced sampling techniques in studying diverse biological systems.

We started with the exploration of a small molecule drug candidate for treating HBV, AT130. To enable accurate simulation of the AT130, we enhanced ffTK to support for the LP particle, a massless virtual entity introduced in CGenFF 4.0 to describe the partially positively charged region found on halogen atoms. The implementation involved modifications to the charge optimization module in ffTK to accommodate charge fitting in the presence of the LP particle, along with updates for the angle and dihedral optimization modules to correctly position the LP particle during angle manipulations. We showcased the new capability of ffTK by optimizing the force field parameters of AT130, which contains a bromine atom, both with and without the LP particle. The comparison of the two parameter sets revealed that the inclusion of the LP particle significantly improved the agreement between quantum mechanical and molecular mechanical water interaction energies, reducing all energy errors to below 1 kcal/mol. In the simulation of AT130 in water, the unoptimized CGenFF parameters failed to reproduce the high flexibility of the dihedrals around the C2–N1 bond as predicted by the quantum mechanical (QM) calculations, regardless of the presence of the LP particle. However, after ffTK parameterization, three and two metastable states of AT130 were sampled when using the parameters with and without the LP particle, respectively. The 25 ns

Y. T. (Andrew) Pang, *Biomolecules' Conformational Changes Studied by Simulations and Enhanced Sampling*, Springer Theses,
https://doi.org/10.1007/978-3-031-70602-8_5

simulations of AT130 in the protein binding pocket told a similar story, showing the molecule spontaneously drifted away from the binding pose in 3 out of 5 replicas of the simulation when using the parameters without optimization. In contrast, all replicas of the simulations using ffTK optimized parameters ended with the ligand root mean square deviation (RMSD) below 4 Å. These results decisively underscore the importance of force field parameterization in accurately capturing the torsion flexibility of small molecules.

Next, we turned our attention to the folding mechanism of the pertactin passenger domain, a highly immunogenic adhesin in *B. pertussis*. The pertactin passenger domain is a large β-helical structure, which is a shared feature among many other autotransporters despite low sequence identities [1]. We began by unfolding the passenger domain using three replicas of steered molecular dynamics (SMD), pulling on the N-terminal end of the passenger domain while holding the C-terminal band fixed. In all three replicas, the passenger domain unfolded strand by strand from the N- to C-terminus, except when unfolding a kink located in the middle of the protein. This unique behavior appears to be due electrostatic interactions between a pair of charged residues around the kink and the "boot"-like geometry of the C-terminal half of the passenger domain. Analyzing the resisting force from the protein during the SMD simulation, one force spike stood out with a peak force of 400 pN. This force spike occurred when unfolding the β-hairpin (V454–S461) in band 1C, which is connected to the neighboring rungs and sits on a large hydrophobic region of the non-β-helical C-terminus. The many hydrogen bonds and strong hydrophobic interactions between the β-hairpin and the surrounding residues explain the high resisting force. This observation aligns with previous experimental findings of a highly stable hydrophobic core at band 1C [2]. Self-learning adaptive umbrella sampling (SLUS) and replica exchange umbrella sampling (REUS) further support these results, showing band 1C as the only band on the C-terminal side that is energetically stable in the folded state when isolated. This suggests the potential role of band 1C in nucleating the folding process of the whole passenger domain. Furthermore, the REUS results demonstrated an energy well of ~10 kcal/mol at the folded state in vectorial folding, contrasting with the unfavorable isolated folding observed. This supports the hypothesis that folding of autotransporter passenger domains only occurs at the outer membrane (OM) due to the specificity of C- to N-terminus secretion.

The final section of this thesis focused on the trimeric spike (S) protein of SARS-CoV-2, which protrudes viral envelope and binds to human angiotensin-converting enzyme 2 (ACE2) receptors. This binding is facilitated by the receptor binding domain (RBD) of the S protein, which transition from a down to up state to enable ACE2 binding. To understand the energetics of RBD opening and the role of glycans, we computed the 2D potential of mean force (PMF) describing this transition for both a fully glycosylated spike and a naked spike using REUS. The simulations. due to the scale of the conformational change and slow convergence, were run for a total of 160 μs on one of the world's fastest supercomputers, Summit at Oak Ridge National Lab (ORNL). Our results indicate that the glycosylated spike favors the down state by ~7 kcal/mol, while the unglycosylated spike shows no statistically significant energy difference between the down and up state. While it might seem advantageous for the virus to have a stronger up-state prescience for increased binding probability, this could also expose it to neutralizing antibodies, negatively impacting the viral survival. Moreover, ACE2 binding also heavily depends on the binding affinity between ACE2 and the S protein, which is likely to be influenced by the presence of glycans [3]. This means that the higher opening energy of the glycosylated spike does not necessarily indicate a higher binding energy. We further analyzed the interactions of individual glycans with various regions of the S protein along the two minimum energy paths (MEPs) obtained, providing a potential explanation for differences in RBD-opening free energies with and without glycans. For the glycosylated up-state spike, the glycans at N122 and N165 disrupt the hydrogen bonds between the opening RBD-A and the neighboring NTD-B, contributing to the higher energy penalty of RBD opening observed in the glycosylated system. With the MEP from the REUS simulation, we performed an antibody accessible surface area (AbASA) analysis with seven well-known antibodies against the S protein. We found that the epitode of BD-368-2 antibody is continuously exposed, explaining its high efficacy.

The enhanced sampling methods discussed in this thesis, while well-established, are not without their limitations. For instance, the REUS simulations we conducted for the SARS-CoV-2 S protein required 1.8 million GPU-hours, a computational cost that may not always be feasible. Consequently, many researchers are actively exploring alternative methods to expand the reach of enhanced sampling. One such promising method is metadynamics [4]. In metadynamics, an artificial bias potential is progressively built during the simulation. This bias, constructed from a series of Gaussian functions deposited in the visited regions of the collective variable space, discourages the system from revisiting previously explored states and effectively fills the energy wells of the system's free energy landscape. Once these energy wells are completely filled and the simulation essentially becomes a random walk, the energy landscape can be recovered as the inverse of the bias potential. Metadynamics offers a unique advantage over umbrella sampling (US) as it does not require the user to provide initial structures across the energy landscape. However, it still necessitates a set of collective variables that defines the meaningful conformational change of the biomolecule. While metadynamics is more cost-effective in terms of computational cost, its accuracy and success heavily depend on the tuning of hyperparameters such as the size of biases and frequency of applying biases. The optimal choice of such these parameters is not as well-explored as in established methods like US, which limits its application when accurate free energy calculations are required. Several variants of metadynamics are being developed to address its limitations, including well-tempered metadynamics [5], bias-exchange metadynamics [6], replica-averaged metadynamics [7], etc. Other enhanced sampling methods aim to obtain not just energetics but also kinetics, such as fluctuation amplification of specific traits (FAST) with Markov state model (MSM) [8] and weighted ensemble [9].

Another significant development in the field pertains to the increasing adoption of machine learning (ML) methods. In recent years, ML, particularly deep learning, has achieved remarkable progress, leading to significant advancements in a wide array of disciplines. Notable applications include natural language processing [10, 11], computer vision [12, 13], and, in the realm of biology, protein structure prediction [14, 15]. A growing trend toward the integration of ML methods is also observed within the domain of enhanced sampling and conformational sampling. For example, Tiwary and his collaborators have introduced the state predictive information bottleneck (SPIB), a method based on variational autoencoder (VAE) that learns to generate the most representative collective variables used to steer enhanced sampling simulations [16, 17]. Similarly, Noe and his colleagues have developed the Boltzmann generator, which has the capability to learn how to replicate the Boltzmann distribution of a biomolecule, even when trained with non-Boltzmann distributed data [18]. It has been suggested that enhanced sampling methodologies can be reformulated to operate within the smoother latent space created by the Boltzmann generator, potentially resulting in substantial improvements in performance.

As a prospective avenue for further research, we are actively engaged in the development of a deep learning model geared towards the swift generation of low-energy transition paths between two specific protein conformations. These generated pathways hold the potential to greatly enhance the efficiency of enhanced sampling techniques and reduce the computational time required to achieve statistical convergence. Our model comprises two integral components: the Explorer and the Structure Builder. The Explorer leverages a progressive growing generative adversarial network (ProGAN) [19] to chart a transition pathway in the residue-residue pairwise distances space, connecting the initial and final conformational states. Meanwhile, the Structure Builder, inspired by the architectural principles of AlphaFold2 [14], is responsible for constructing the all-atom protein structure based on the output generated by the Explorer. To fine-tune the model's ability to produce low-energy, biophysically plausible protein transition pathways, we employ an iterative approach that combines active learning and energy validation using a molecular dynamics (MD) energy minimizer. Currently, we are actively refining the details of the network architecture and the training procedure. Once this model reaches a mature stage, it can be applied to the proteins under investigation in this thesis, specifically the pertactin passenger domain and the SARS-CoV-2 S protein. These proteins will serve as both validation subjects for the model's outputs and as avenues for further exploration of potential transition pathways.

On the individual systems examined within the scope of this thesis, there remains a trove of captivating aspects yet to be explored. In the case of HBV capsid assembly modulators, the parameterization of AT130 provides a foundational platform for delving deeper into its interactions with the capsid proteins, which is also undergo substantial conformational changes during assembly. Moreover, the updated ffTK also allow us to go beyond AT130 and parameterize other halogenated molecules, including other HBV capsid assembly modulators. Indeed, our laboratory has embarked on further investigations, focusing on how AT130 and sixteen additional capsid assembly modulators influence the conformation of the HBV capsid assembly intermediate. Intriguingly, our studies have revealed that distinct conformations are induced, contingent upon whether the bound modulator accelerates or misdirects the assembly process [20]. Nonetheless, several crucial questions still await answers, particularly concerning how these capsid assembly modulators precisely impact the capsid assembly process. To address these queries, the employment of additional free energy calculations using REUS or metadynamics can play a pivotal role in quantifying the diverse effects exerted by various modulators. Such investigations promise to shed light on the intricacies of capsid assembly and further our understanding of how these modulators can be harnessed to manipulate this process effectively.

In the context of pertactin passenger domain secretion, future experiments could further validate the role of the hydrophobic core at the C-terminus. This could be achieved by relocating the unfolding-resistant β-hairpin to other parts of the passenger domain sequence to assess whether it maintains the same folding stability. Additionally, the construction of a chimera autotransporter, which combines the barrel and passenger domain from two different classical autotransporters, could help ascertain whether the barrel plays a more active role than simply forming a conduit for the passenger domain to traverse. In addition, our investigations have thus far been concentrated on evaluating the isolated and vectorial folding free energy of a few selected termini bands. To gain a more comprehensive understanding of this process and address the fundamental question of whether there exists an energy gradient governing folding from the C- to N-terminus [21], an extension of our calculations to encompass a wider range of bands spanning the passenger domain is warranted. This expanded analysis can provide a holistic view of the energy landscape associated with folding in this domain. Moreover, our SMD and SLUS simulations have primarily focused on exploring the folding pathway from the C- to N-terminus. A pivotal step in strengthening our conclusions and validating the hypothesis that folding predominantly occurs from the C- to N-terminus is to undertake a complementary investigation into the reverse folding process, specifically from the N- to C-terminus. Such a complementary study can provide a more comprehensive perspective on the folding dynamics of the pertactin passenger domain. Finally, a more realistic simulation setting is needed to evaluate the effects of the autotransporter β-barrel and the extracellular environment on the folding of the passenger domain. Experimental studies have shown that the extracellular loop of the translocator β-barrel of the Type Va autotransporter BrkA forms a hydrophobic surface that is essential for passenger domain secretion [22]. More recently, another group has shown similar results for another Type Va autotransporter Pet [23]. Both works suggest that the folding of the passenger domain may be nucleated via β-augmentation on top of the hydrophobic surface of the translocator's extracellular loop. To scrutinize and validate this proposition, it is imperative to conduct equilibrium and SMD simulations of pertactin in the OM, thus closely mimicking the authentic cellular environment. Such simulations will provide insights into the dynamics of passenger domain folding in a biologically relevant context.

Regarding the SARS-CoV-2 S protein, our study was conducted on the original strain of the virus from 2019. Since that time, numerous new variants have emerged, with a few Omicron subvariants currently dominating [24]. Investigating how different mutations impact the energetics of RBD opening can significantly enhance our understanding of the SARS-CoV-2 infection mechanism. The research will contribute to the ongoing battle against the COVID pandemic and help us prepare for potential future outbreaks stemming from other coronaviruses. Additionally, the experience gained from running large-scale REUS simulations and managing terabytes of data has led to the development of valuable job preparation and data analysis scripts. These tools empower us to conduct enhanced sampling simulations of even more complex systems in the future. One such system on our horizon is the bacterial Type IV pilus (T4P). T4P serves as a critical virulence factor, with diverse functions ranging from cellular adherence to DNA uptake from the environment. Our forthcoming research aims to uncover the molecular intricacies governing T4P growth and retraction. By employing an extensive MD simulation model encompassing over 5 million atoms, we will simulate the Type IV competence pilus machine within a realistic *Vibrio Cholerae* membrane using REUS simulations. This exploration of the energy landscape governing pilus movement will allow us to identify potential checkpoints and barriers, providing valuable insights for the design of therapeutic interventions.

References

1. M. Junker et al., Pertactin β-helix folding mechanism suggests common themes for the secretion and folding of autotransporter proteins. Proc. Natl. Acad. Sci. U. S. A. **103**(13), 4918–4923 (2006)
2. J.P. Renn, P.L. Clark, A conserved stable core structure in the passenger domain β-helix of autotransporter virulence proteins. Biopolymers **89**(5), 420–427 (2008)
3. A. Acharya, D.L. Lynch, A. Pavlova, Y.T. Pang, J. Gumbart, ACE2 glycans preferentially interact with SARS-CoV-2 over SARS-CoV. Chem. Commun. **57**, 5949–5952 (2021)
4. A. Laio, M. Parrinello, Escaping free-energy minima. Proc. Natl. Acad. Sci. U. S. A. **99**(20), 12562–12566 (2002)
5. A. Barducci, G. Bussi, M. Parrinello, Well-tempered metadynamics: a smoothly converging and tunable free-energy method. Phys. Rev. Lett. **100**(2), 020603 (2008)
6. S. Piana, A. Laio, A bias-exchange approach to protein folding. J. Phys. Chem. B **111**(17), 4553–4559 (2007)
7. C. Camilloni, A. Cavalli, M. Vendruscolo, Replica-averaged metadynamics. J.Chem. Theory Comput. **9**(12), 5610–5617 (2013)
8. M.I. Zimmerman, G.R. Bowman, FAST conformational searches by balancing exploration/exploitation trade-offs. J. Chem. Theory Comput. **11**(12), 5747–5757 (2015)
9. D. Bhatt, B.W. Zhang, D.M. Zuckerman, Steady-state simulations using weighted ensemble path sampling. J. Chem. Phys. **133**(1), 014110 (2010)
10. T. Brown et al., Language models are few-shot learners, in *Adv. Neural Inf. Process.*, ed. by H. Larochelle, M. Ranzato, R. Hadsell, M. Balcan, H. Lin, vol. 33 (Curran Associates, 2020), pp. 1877–1901
11. OpenAI, *GPT-4 technical report* (2023). arXiv: 2303.08774 [cs.CL]
12. R. Rombach, A. Blattmann, D. Lorenz, P. Esser, B. Ommer, High-resolution image synthesis with latent diffusion models, in *Proc. IEEE Comput. Soc. Conf. Comput. Vis. Pattern Recognit.* (2022), pp. 10684–10695

13. A. Ramesh, P. Dhariwal, A. Nichol, C. Chu, M. Chen, Hierarchical text-conditional image generation with clip latents. Preprint. arXiv:2204.06125 **1**(2), 3 (2022)
14. J. Jumper et al., Highly accurate protein structure prediction with AlphaFold. Nature **596**(7873), 583–589 (2021)
15. M. Baek et al., Accurate prediction of protein structures and interactions using a three-track neural network. Science **373**(6557), 871–876 (2021)
16. D. Wang, P. Tiwary, State predictive information bottleneck. J. Chem. Phys. **154**(13), 134111 (2021)
17. S. Mehdi, D. Wang, S. Pant, P. Tiwary, Accelerating all-atom simulations and gaining mechanistic understanding of biophysical systems through state predictive information bottleneck. J. Chem. Theory Comput. **18**(5), 3231–3238 (2022)
18. F. Noé, S. Olsson, J. Köhler, H. Wu, Boltzmann generators: Sampling equilibrium states of many-body systems with deep learning. Science **365**(6457), eaaw1147 (2019)
19. T. Karras, T. Aila, S. Laine, J. Lehtinen, Progressive growing of GANs for improved quality, stability, and variation, in *Int. Conf. Learn. Represent.* (2018)
20. A. Pavlova et al., The mechanism of action of hepatitis B virus capsid assembly modulators can be predicted from binding to early assembly intermediates. J. Med. Chem. **65**(6), 4854–4864 (2022)
21. M. Junker, R.N. Besingi, P.L. Clark, Vectorial transport and folding of an autotransporter virulence protein during outer membrane secretion. Mol. Microbiol. **71**(5), 1323–1332 (2009)
22. Y. Zhai et al., Autotransporter passenger domain secretion requires a hydrophobic cavity at the extracellular entrance of the β-domain pore. Biochem. J. **435**, 577–587 (2011)
23. X. Yuan et al., Molecular basis for the folding of β-helical autotransporter passenger domains. Nat. Commun. **9**, 1395 (2018)
24. Centers for Disease Control and Prevention, *COVID data tracker* (2023). https://covid.cdc.gov/covid-data-tracker. Accessed on October 24, 2023, Atlanta, GA

Appendix A
Supporting Information: Parameterization of a Drug Molecule with a Halogen σ-Hole Particle Using ffTK: Implementation, Testing, and Comparison

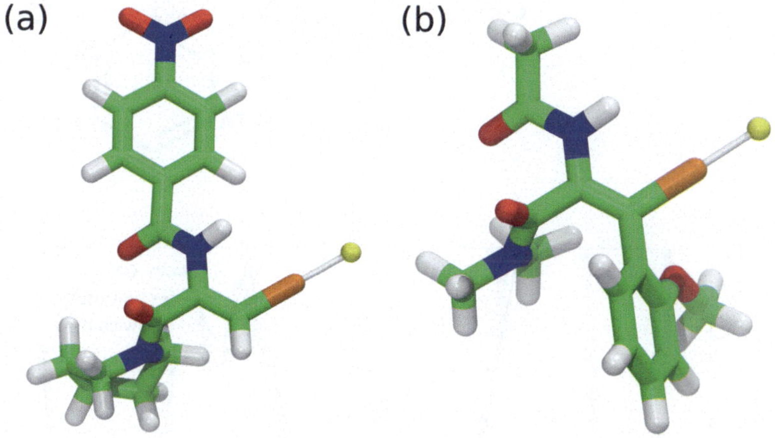

Fig. A.1 Reduced structures for (**a**) partial charges optimization and (**b**) bonds and angles optimization

Y. T. (Andrew) Pang, *Biomolecules' Conformational Changes Studied by Simulations and Enhanced Sampling*, Springer Theses,
https://doi.org/10.1007/978-3-031-70602-8

Fig. A.2 Atom names of all heavy atoms

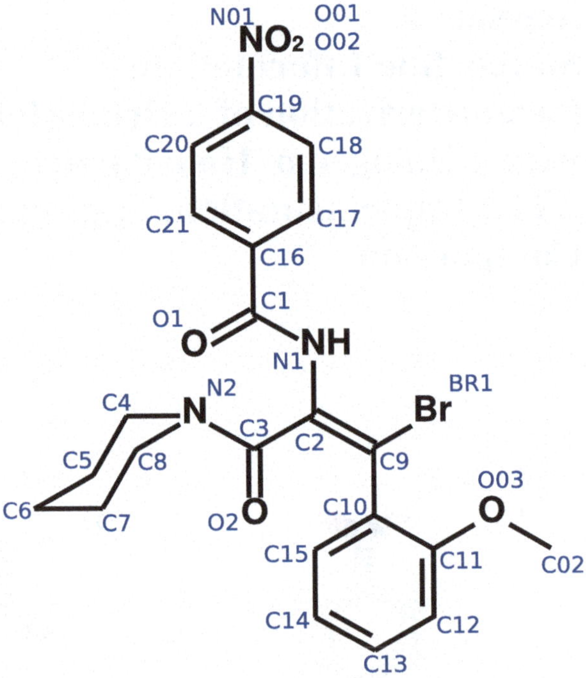

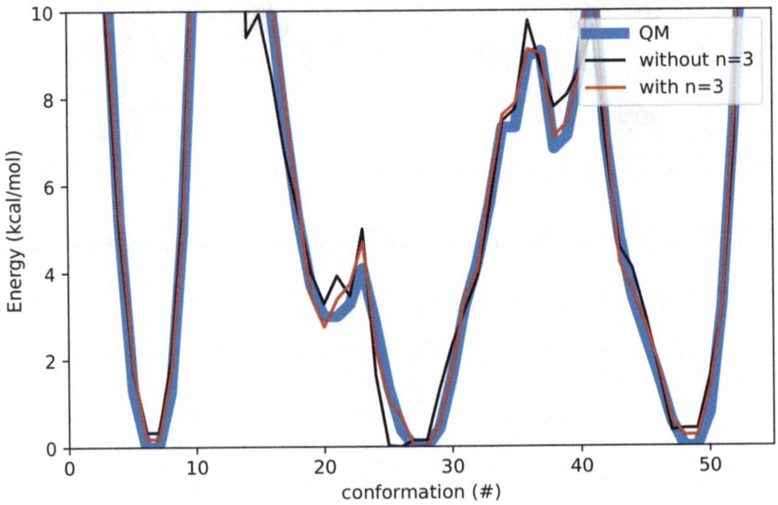

Fig. A.3 Comparing the PES with and without the n = 3 multiplicity for dihedral CG2DC1 CG2D1O NG2S1 CG2O1 (C9-C2-N1-C1)

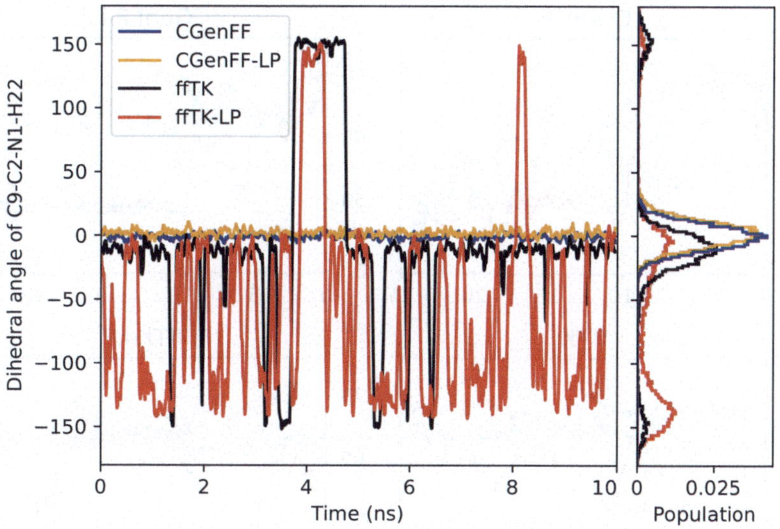

Fig. A.4 The dihedral angle of C9-C2-N1-H22 trajectory and populations in a 10-ns simulation in water with different parameters before reoptimizing the dihedral parameters around the bond C-N

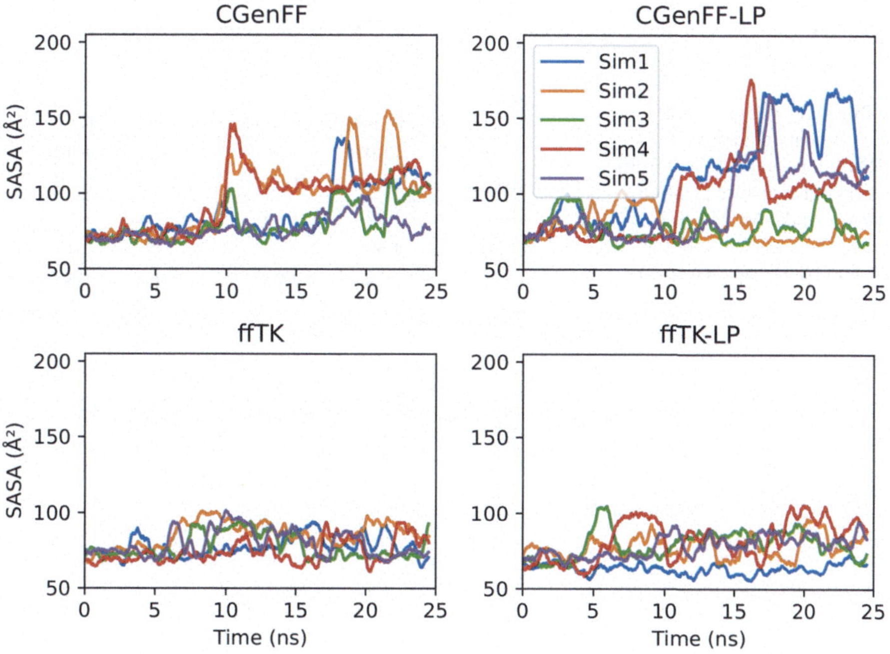

Fig. A.5 SASA of AT130 over time

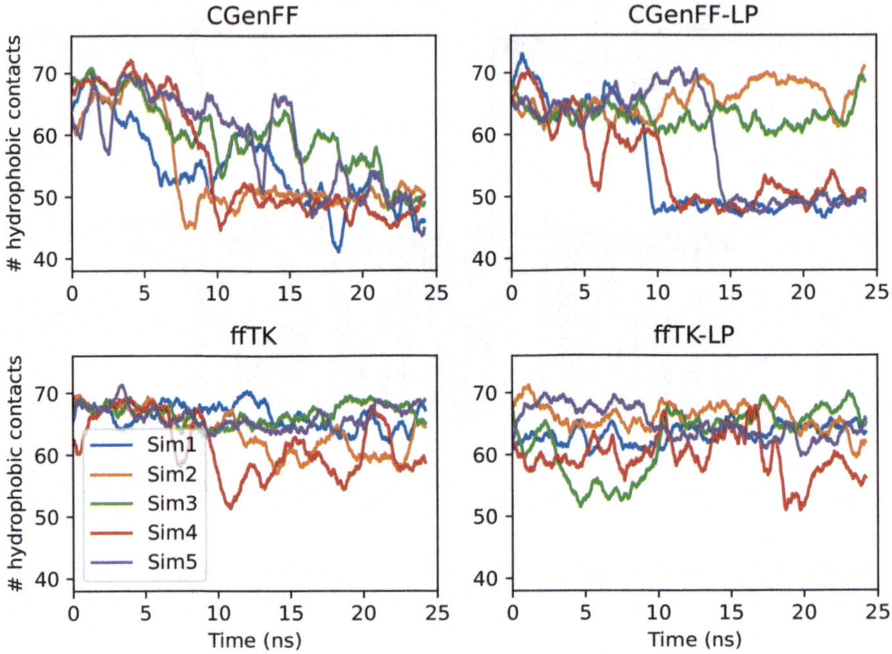

Fig. A.6 Number of hydrophobic contacts between AT130 and the protein complex over time

Table A.1 Average root mean square deviation (RMSD, in Å) of AT130 from its crystal-structure position when simulating AT130 in the binding pocket with the protein backbone restrained

Simulation	CGenFF	CGenFF-LP	ffTK	ffTK-LP
Sim1	4.60	5.36	1.57	1.39
Sim2	4.82	1.51	2.11	1.60
Sim3	3.64	1.53	1.40	1.67
Sim4	4.59	4.62	1.34	2.43
Sim5	2.28	4.95	1.45	1.59

Appendix B
Supporting Information: Uncovering the Folding Mechanism of Pertactin: A Comparative Study of Isolated and Vectorial Folding

B.1 Analysis of the SMD Simulations

Other than the major force spikes as discussed in the main text, we describe here the investigation of all other spikes, and correlate them to the unfolding of each band. Spikes at pulled distances of 15, 22, 30–33, 40, and 50 nm represent the unfolding of bands 3N, 4N, 5N, 6N and 7N, respectively. Following these is a relatively chaotic region. Because all the folded structures that the 35-residue-long loop linking rungs 9N and 10N (loop 9N) was originally attached have fully unfolded, this loop becomes loosely attached to any exposed region of the protein while band 8N is being pulled apart. This creates a number of minor force spikes until band 8N is fully unfolded at a pulled distance of 60 nm. After band 9N unfolds at 63 nm, the resistance force remains low as loop 9N being extended. A force spike then appears at 73 nm after loop 9N is fully extended and band 10N started to unfold.

Continuing at the C-terminal side after the kink unfolds, we see a series of force spikes at 87, 92, 96–100, 110, 116, 120, and 126 nm, which arise from the unfolding of bands 7C, 6C, 5C, 4C, 3C, 2C, and 1C, respectively. Finally, we reach the largest force spike at $\sim$130 nm from unfolding the β-hairpin structure at band1C as described in the main text. The last force spike at 136 nm is due to pulling the last C-terminal rung away from the non-helical C-terminus.

Y. T. (Andrew) Pang, *Biomolecules' Conformational Changes Studied by Simulations and Enhanced Sampling*, Springer Theses,
https://doi.org/10.1007/978-3-031-70602-8

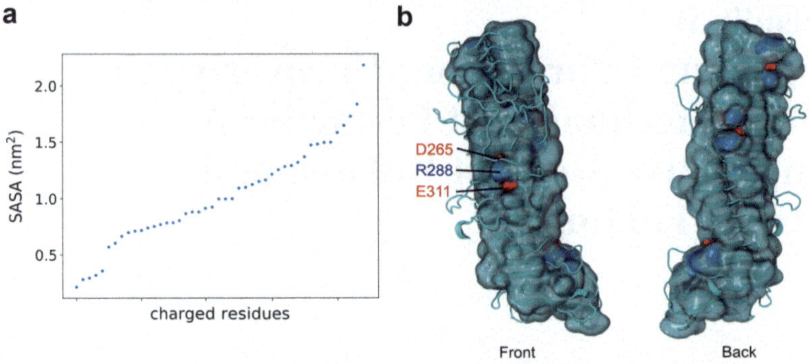

Fig. B.1 Exposure of charged residues to solvent. (**a**) SASA of charged residues that are part of a β-strand. The five residues with the lowest SASA are D265, D197, D426, D311, and D53, in ascending order of their SASAs. (**b**) The location of the five-lowest-SASA residues on the β-helix together with the nearby residues are highlighted as red (positively charged) and blue (negatively charged) spheres. The surface of the protein, excluding the five-lowest-SASA residues, is shown in cyan

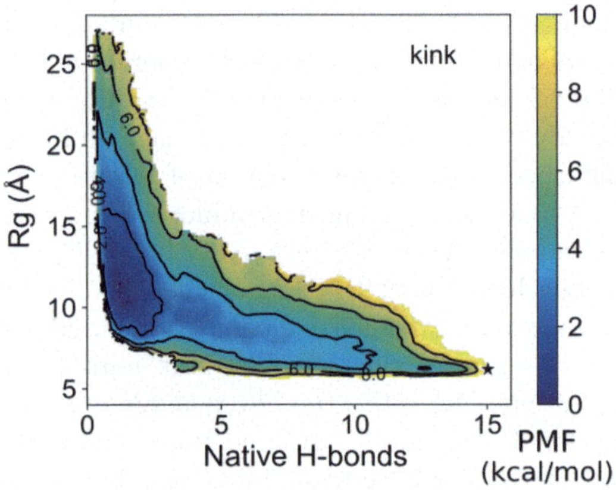

Fig. B.2 2D PMF for isolated folding of the kinked band along two collective variables, N_{hb} and R_{G}. Native structure coordinates indicated with a star. Our SMD simulations indicated that the kinked band is more stable when folded on top of the folded C-terminus compared to some C-terminal bands such as 5C and 7C. However, the native structure of the kinked band is not at an energy minimum when folded in isolation

Table B.1 Residues of fixed C_α atoms in the SMD simulations

System	Residues of fixed C_α atoms
N-terminus to rung 10N (D1–G284)	207–212, 216–225, 262–267, and 270–278
rung 8N to rung 5C (A184–G391)	294–298, 300–312, 325–330, 334–343, 349–354, 358–364, 379–384, 388, and 389
rung 10N to the C-terminus (Y262–P539)	483–539
rung 4C to the C-terminus (V396–P539)	483–539

Table B.2 Isolated bands of the pertactin β-helical passenger domain

Fragment	Residues	Native HBonds	Initial # of windows	Final # of windows	SLUS iterations
1N	1–38	7	187	414	17
2N	16–64	16	234	769	14
3N	45–96	17	269	698	14
kink	262–313	18	267	816	9
3C	396–428	11	226	560	10
2C	411–452	12	293	542	9
1C	438–481	16	252	589	9

Appendix C
Supporting Information: SARS-CoV-2 Spike Opening Dynamics and Energetics Reveal the Individual Roles of Glycans and Their Collective Impact

C.1 Analysis

Minimum Energy Path (MEP) The MEPs were computed using the algorithm by Ensing et al. [1] and the code implemented by Mahmoud Moradi [2]. The two energy minima and the saddle point (the point with lowest energy to cross an energy barrier) were found using a separate in-house python script, and the MEP from the saddle point to each energy minimum was found using the code by Moradi. The paths were smoothed by the least-square fitting B-spline function from python library SciPy [3].

Mean First Passage Time (MFPT) Let $\lambda(\mathbf{d}, \boldsymbol{\theta})$ be the one-dimensional path parameter that describes the position of the system along the MEP. Assuming the dynamics along λ can be effectively described by a diffusive model, we may apply the Smoluchowski diffusion equation to describe the process [4, 5]:

$$\frac{\partial}{\partial t} p(\lambda, t|\lambda_0, 0) = \frac{\partial}{\partial \lambda} D(\lambda) e^{-\beta F(\lambda)} \frac{\partial}{\partial \lambda} \left(e^{\beta F(\lambda)} p(\lambda, t|\lambda_0, 0) \right) \tag{C.1}$$

where $p(\lambda, t|\lambda_0, 0)$ is the probability of finding the system at λ after time t, given it was at λ_0 at time 0, $D(\lambda)$ is a position-dependent diffusion constant, and $F(\lambda)$ is the free energy at λ. By rearranging the terms, the MFPT $\bar{\tau}_{FP}$, or the rate inverse k^{-1}, from the initial (A) to the final (B) state is given by:

$$k^{-1} = \bar{\tau}_{FP} = \int_{\lambda_A}^{\lambda_B} \frac{1}{D(\lambda)} e^{\beta F(\lambda)} \int_{\lambda_0}^{\lambda} e^{-\beta F(\lambda')} \mathrm{d}\lambda' \mathrm{d}\lambda \tag{C.2}$$

While $F(\lambda)$ was readily available from the PMF obtained through REUS, $D(\lambda)$ was approximated using a generalized-Langevin-equation-based method, derived by Roux and co-workers and implemented by Gaalswyk et al. [6]. From a time

Y. T. (Andrew) Pang, *Biomolecules' Conformational Changes Studied by Simulations and Enhanced Sampling*, Springer Theses,
https://doi.org/10.1007/978-3-031-70602-8

series of a coordinate x_k with the system simulated under a harmonic restraint, the method computes the diffusion constant D_k along x_k by relating it to its velocity autocorrelation function (VACF). With a series of coordinate transformations [7], D_k is transformed from the Cartesian space to the collective variable space (D_{ij}), and then to the path variable space ($D(\lambda)$), which was then inserted back into Eq. (C.2).

$$D_{ij} = \sum_k D_k \left\langle \frac{\partial z_i}{\partial x_k} \frac{\partial z_j}{\partial x_k} \right\rangle, \; D(\lambda) = \sum_{ij} D_{ij} \frac{\partial \lambda}{\partial z_i} \frac{\partial \lambda}{\partial z_j} \tag{C.3}$$

To determine D_k for each atom and each of its coordinates involved in the definition of the collective variables d and ϕ, we ran a 1-ns simulation for each window along the MEP, with all C_α atoms of the protein restrained by a force constant of 5 kcal mol^{-1}Å^{-2}. A 2-fs time step was used without the application of HMR. Other simulation parameters were identical to the REUS simulations.

Kinetics Analysis We modeled the down-to-up transition and subsequent binding of the RBD to ACE2 according to the chemical equation

$$\mathrm{D + ACE2} \underset{k_{\mathrm{close}}}{\overset{k_{\mathrm{open}}}{\rightleftharpoons}} \mathrm{U + ACE2} \underset{k_{\mathrm{off}}}{\overset{k_{\mathrm{on}}}{\rightleftharpoons}} \mathrm{U{:}ACE2}$$

The associated master equation is

$$\frac{\mathrm{d[D]}}{\mathrm{d}t} = -k_{\mathrm{open}}[\mathrm{D}] + k_{\mathrm{close}}[\mathrm{U}] \tag{C.4}$$

$$\frac{\mathrm{d[ACE2]}}{\mathrm{d}t} = k_{\mathrm{off}}[\mathrm{U{:}ACE2}] - k_{\mathrm{on}}[\mathrm{U}][\mathrm{ACE2}] \tag{C.5}$$

$$\frac{\mathrm{d[U]}}{\mathrm{d}t} = k_{\mathrm{open}}[\mathrm{D}] - k_{\mathrm{close}}[\mathrm{U}] - k_{\mathrm{on}}[\mathrm{U}][\mathrm{ACE2}] + k_{\mathrm{off}}[\mathrm{U{:}ACE2}] \tag{C.6}$$

$$\frac{\mathrm{d[U{:}ACE2]}}{\mathrm{d}t} = -k_{\mathrm{off}}[\mathrm{U{:}ACE2}] + k_{\mathrm{on}}[\mathrm{U}][\mathrm{ACE2}] \tag{C.7}$$

The rates k_{open} and k_{close} come from our own calculations. For the glycosylated spike, $k_{\mathrm{open}} = 0.68$/s and $k_{\mathrm{close}} = 1.17 \times 10^3$/s, while for the un-glycosylated spike, $k_{\mathrm{open}} = 7.01 \times 10^3$/s and $k_{\mathrm{close}} = 2.01 \times 10^3$/s. The rates $k_{\mathrm{on}} = 1.40 \times 10^6$/M·s and $k_{\mathrm{off}} = 6.54 \times 10^{-3}$/s were taken from Lan et al. [8] for the RBD alone in order to isolate the effects of binding from conformational changes in the spike. The system of first-order ordinary differential equations was numerically solved up to 250 s, for which all systems reached steady-state populations, using Mathematica along with the initial conditions $[\mathrm{D}]_i$ = 1 nM, $[\mathrm{U}]_i$ = 0 nM, $[\mathrm{U{:}ACE2}]_i$ = 0 nM, and $[\mathrm{ACE2}]_i$ = 15 nM. The concentration chosen for ACE2 is the same order of magnitude as $K_{\mathrm{D}} = 4.67$ nM, and we note that the qualitative results did not change for different values of $[\mathrm{ACE2}]_i$ although absolute populations did shift.

Hydrogen Bond Analysis The number of hydrogen bonds formed between RBD-A and other domains of the spike in the REUS trajectories were measured using the

HBonds Plugin of VMD. When the distance between a hydrogen-bond donor atom (D) and an acceptor atom (A) is below 3.5 Å and the angle D-H-A is less than 35° from 180°, a hydrogen bond was considered to be formed between D and A. The d-ϕ space is broken into small bins and the number of hydrogen bonds was averaged over each bin.

Contact Analysis Contact calculations were performed using a cutoff distance of 3.5 Å. In other words, when two heavy atoms from two different selections come within 3.5 Å, we count that as one contact. For the analysis done in Fig. 4.4c-d and Fig. C.8, the d-ϕ space is broken into small bins and the number of contacts was averaged over each bin. For the analysis done in Fig. C.9, we separated conformations from the MEP in two categories and the number of contacts were

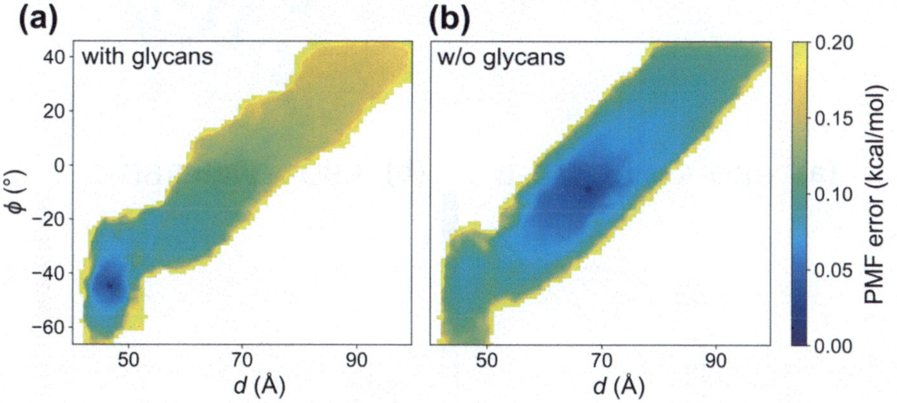

Fig. C.1 Uncertainties of the free energy differences with the lowest point on the PMF computed by MBAR for the (**a**) glycosylated and (**b**) un-glycosylated system

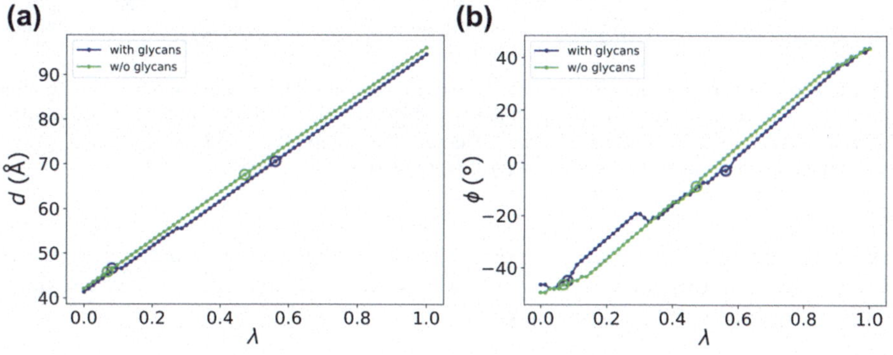

Fig. C.2 Path parameter λ. (**a, b**) The path parameter is defined as a value going from zero to one along the MEP from the bottom-left to the top-right corner of the PMF. The corresponding d and ϕ are plotted against λ in (**a**) and (**b**), respectively. The position of the down- and up-state energy minima are marked as empty circles

Fig. C.3 Snapshot from the un-glycosylated REUS simulation showing the β-sheets of RBD-A (pink) and NTD-B (blue) align with each other and form strong hydrogen bonds when the RBD-A is in the up state

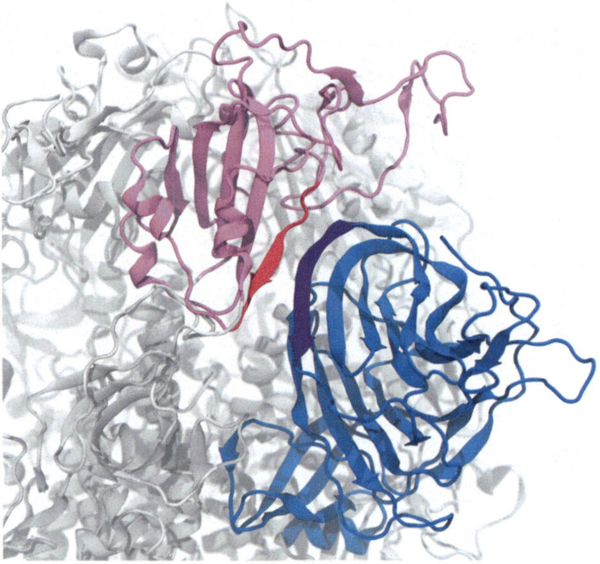

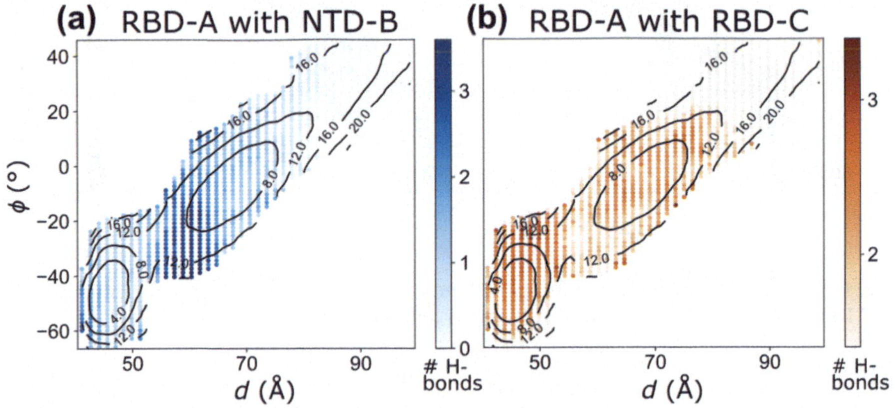

Fig. C.4 Hydrogen bond analysis for the glycosylated system. The average number of hydrogen bonds formed (**a**) between RBD-A and NTD-B and (**b**) between RBD-A and RBD-C are plotted against the two collective variables, d and ϕ. Contour lines of the PMF of the glycosylated system are plotted on top to show the location of the energy barrier

averaged separately for each category. The two categories were defined by the d values. The conformations with $d \leq 54.9$ Å were considered as the down state, while the rest were taken as the up state.

Antibody Accessible Surface Area (AbASA) AbASA calculations were performed using the solvent-accessible surface area measurement tool as implemented in VMD with a 7-Å probe for each frame along the MEP. For the glycosylated

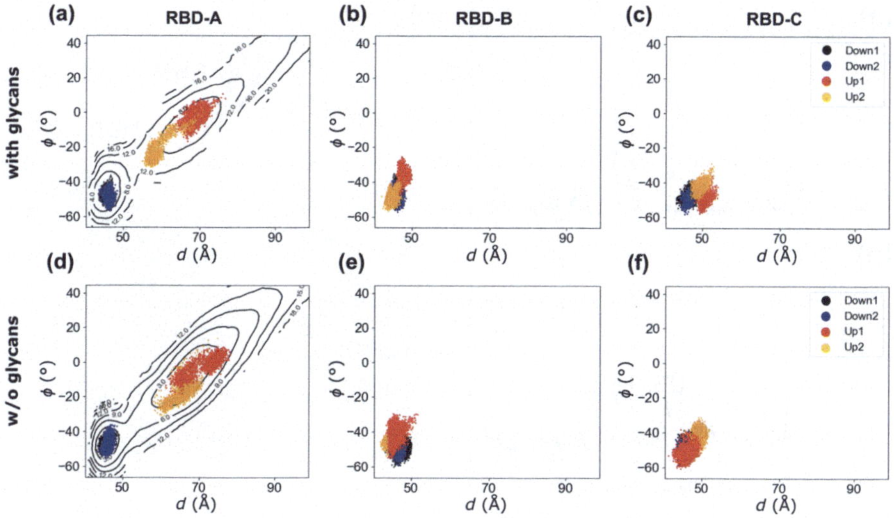

Fig. C.5 Collective variable d vs. ϕ for the RBDs in 2 × 2-μs equilibrium simulations. (**a–c**) Glycosylated system. (**d–f**) Un-glycosylated system. The S-protein protomers in the down state, including (**a, d**) RBD-A that started at the down state (black/blue), (**b, e**) RBD-B, and (**c, f**) RBD-C, all remained in the free energy well of the down state for all replicas, with or without glycans. The up-state RBD-A (red/yellow) was less stable in comparison

Fig. C.6 Contact between glycans at N165 and N343 along the MEP

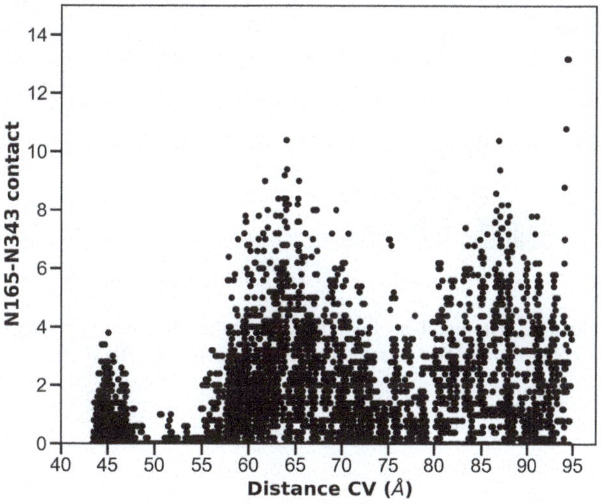

system, we performed two accessible surface area calculations for each frame around each epitope using the *-restrict* option: AbASA of (calculation #1) protein and (calculation #2) protein + glycans. The difference in the accessible surface area obtained from these two calculations gives the coverage provided by the glycans only.

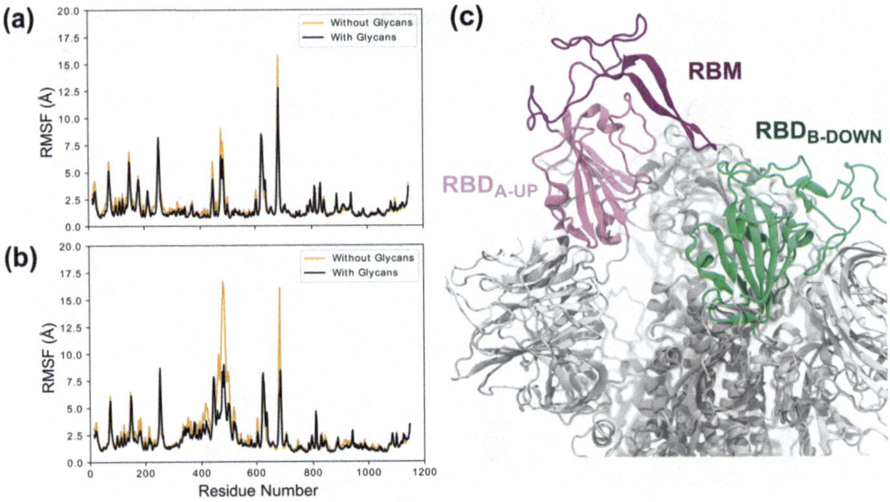

Fig. C.7 Root-mean-square fluctuation (RMSF) for equilibrium simulations. (**a**) RMSF of the down protomers, averaged over the 3 protomers and 2 replicas. (**b**) RMSF of the single up protomer, averaged over both replicas. RMSF computed after alignment and averaging over all C_α atoms of the trimer. (**c**) Snapshot from the equilibrium simulation without glycans, showing the RBM region (438–506) of the up-state protomer attached to the neighboring RBD-B

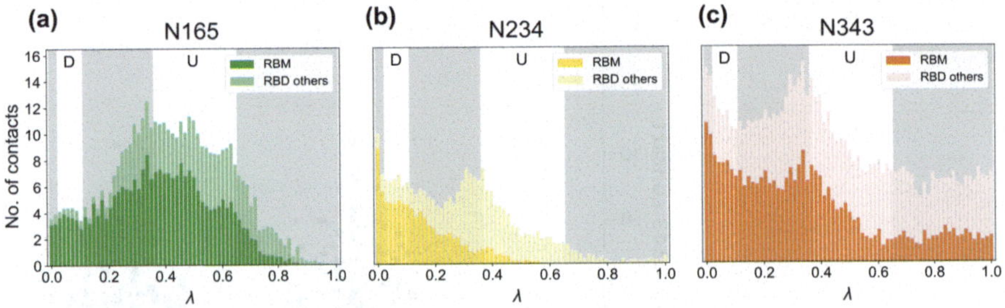

Fig. C.8 The number of contacts along the MEP between the glycans at (**a**) N165, (**b**) N234 and (**c**) N343, and RBD-A, which is separated into the RBM (439–506) and the non-RBM

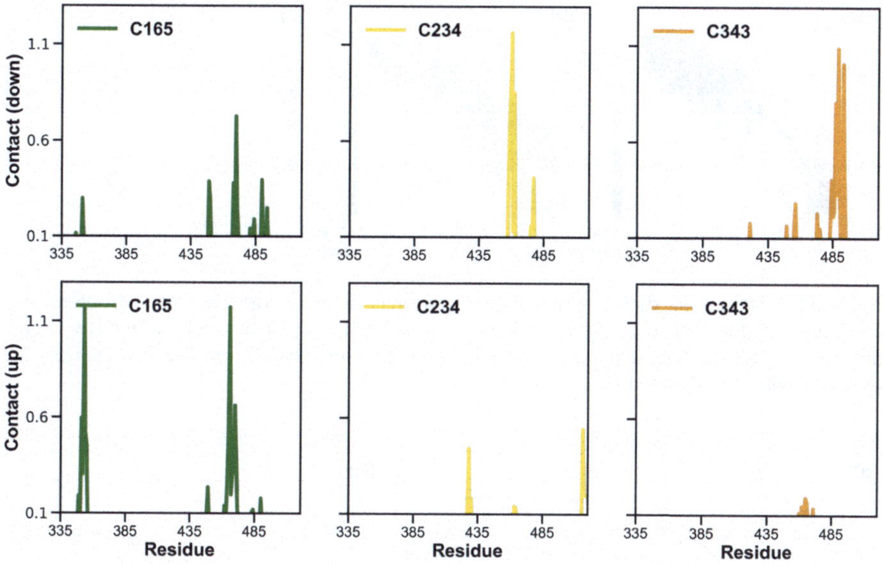

Fig. C.9 Average contact between glycans and RBD residues in up and down conformations of the MEP. Conformations in the range 43.1 Å $\leq d \leq$ 48.9 Å were considered as the down state, while the conformations with 60.1 Å $\leq d \leq$ 75.6 Å were taken as the up state

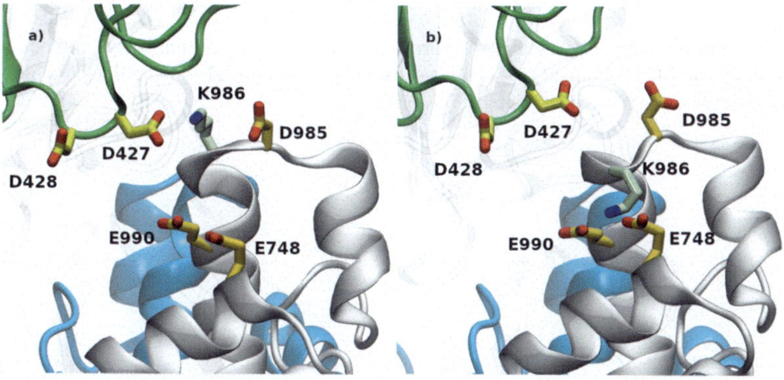

Fig. C.10 Intra-protomer and inter-protomer K986 interactions. K986 transiently interacts with (**a**) D427 from a neighboring protomer or (**b**) D985, E748, and E990 from the same one. D427, D428, and the green ribbon are from protomer B while K986, D985, E748, E990, as well as the silver and blue ribbons, are from protomer A

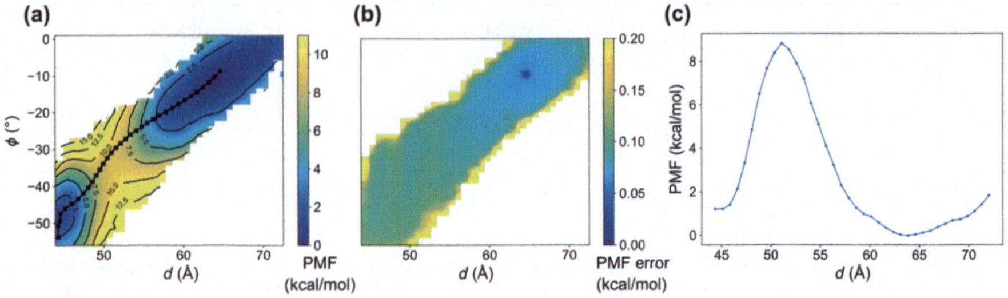

Fig. C.11 PMF of the diproline-mutated un-glycosylated spike. (**a**) The 2D PMF along two collective variables, d and ϕ. The black lines show the MEP. (**b**) Uncertainties of the free energy differences with the lowest point on the PMF computed by MBAR. (**c**) The free energies are projected onto d and plotted as an 1D PMF

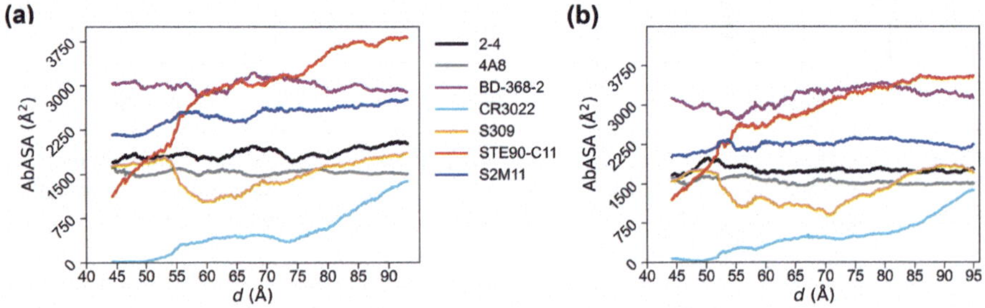

Fig. C.12 Epitope analysis of selected antibodies. Exposed area on antibody epitopes without glycans along the MEP obtained (**a**) with glycans and (**b**) without glycans. All accessible surface area calculations were performed using a 7-Å probe

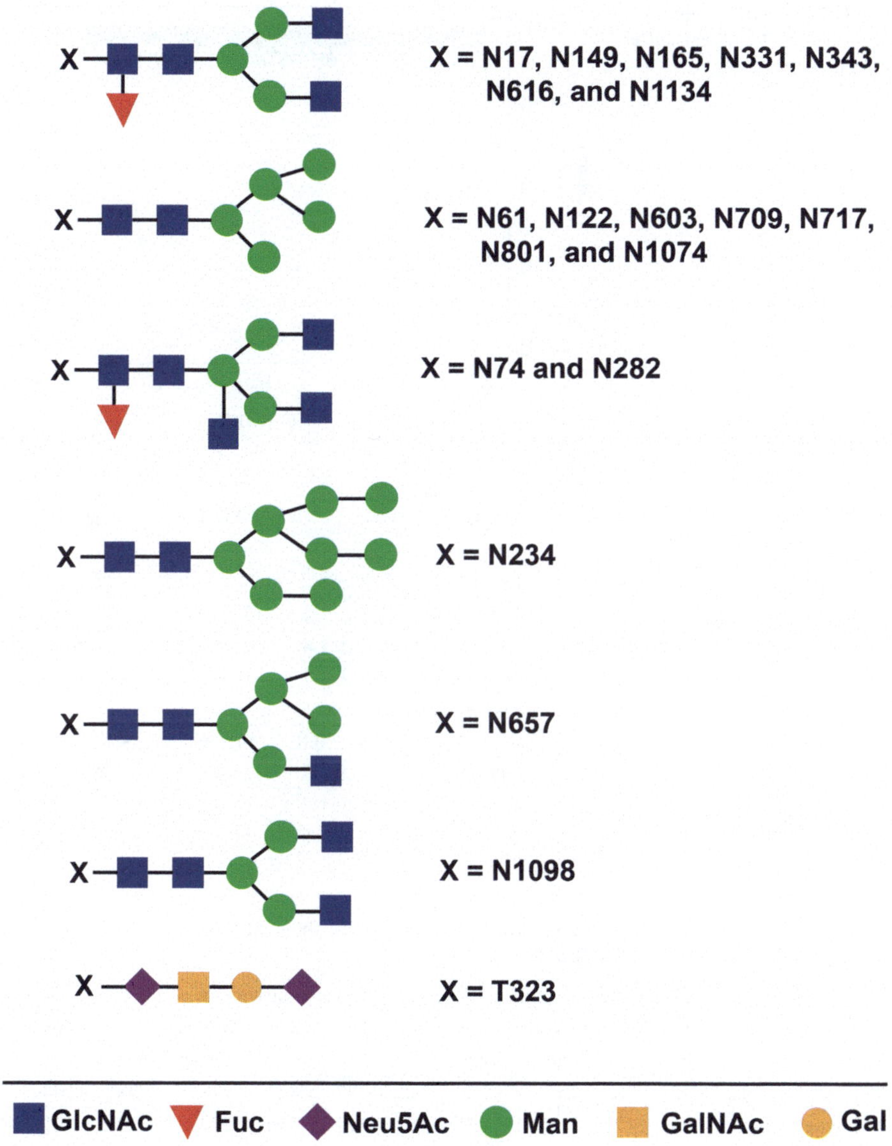

Fig. C.13 Glycan compositions used in this study

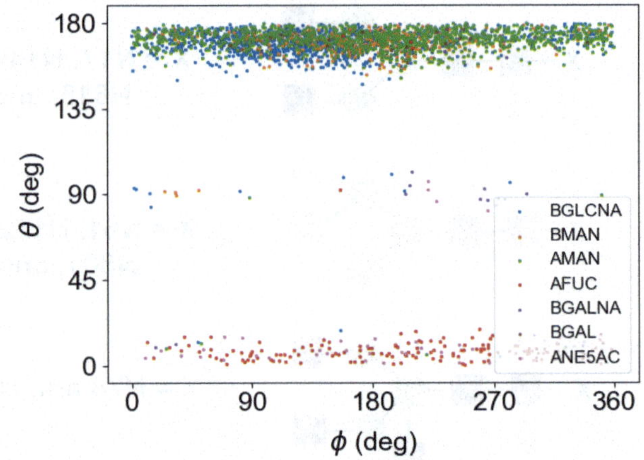

Fig. C.14 Cremer-Pople angles [9] of the glycans after simulated annealing

Table C.1 The average position of the stationary domains used to define the two collective variables throughout the REUS simulations and their standard deviations

Domains	x (Å)	y (Å)	z (Å)
Glycosylated			
SD1-B	−31.6 ± 1.1	14.1 ± 1.1	−13.1 ± 1.0
SD1-A	2.2 ± 1.4	−34.2 ± 1.1	−11.8 ± 1.6
SD2-A	26.5 ± 1.4	−25.4 ± 1.6	11.6 ± 1.2
NTD-A	45.8 ± 1.7	−11.0 ± 1.5	−21.8 ± 0.9
Unglycosylated			
SD1-B	−31.4 ± 0.8	14.7 ± 1.2	−12.4 ± 1.0
SD1-A	2.6 ± 1.7	−34.8 ± 1.2	−10.9 ± 2.1
SD2-A	25.6 ± 1.2	−25.3 ± 1.5	12.5 ± 1.0
NTD-A	45.2 ± 1.6	−11.2 ± 1.6	−21.7 ± 1.1

Table C.2 The PMF cutoffs for a window to be retained for the next stage of REUS simulation and the resulting number of REUS windows

	PMF cutoff (kcal/mol)					Number of windows				
	1	2	3		4	1	2	3		4
Round			Down	Up				Down	Up	
WT with glycans	30	28	25	26	∞	387	422	178	700	1049
WT w/o glycans	24	12	25	25	∞	392	356	257	669	1211
diproline mutant w/o glycans	28	–	–	–	–	353	–	–	–	–

Table C.3 Details of the antibodies investigated in this study. Antibodies 2–4 [10], 4A8 [11], and BD-368-2 [12] were obtained from B cells of patients after SARS-CoV-2 infection, while S309 [13] was extracted from B cells of SARS-CoV infected patients. Antibodies CR3022 [14] and STE90-C11 [15] were created from phage libraries

Antibody	Binds to	Residues in epitope	PDB id	Reference
2–4	SARS-CoV-2 RBD	446–447, 449, 452–453, 455–456, 483–487, 489, 490, 492–496, 498	6XEY	Liu et al. [10]
4A8	SARS-CoV-2 NTD	143–148, 150–152, 158, 245–251, 256–257	7C2L	Chi et al. [11]
BD-368-2	SARS-CoV-2 RBD	346, 351, 444–447, 449–450, 452, 470, 472, 478–486, 490, 492, 494, 498	7CHH	Du et al. [12]
CR3022	SARS-CoV-2 RBD	369–372, 374–386, 389–390, 392, 427–430, 515–517, 519	6W41	Yuan et al. [14]
S309	SARS-CoV/SARS-CoV-2 RBDs	333–335, 337, 339–341, 343–346, 354, 356–361, 440, 441, 444, 509	6WPS	Pinto et al. [13]
STE90-C11	SARS-CoV-2 RBD	403, 405, 406, 408, 409, 415–417, 420, 421, 449, 453, 455–460, 473–477, 486–487, 489, 493–496, 498, 500–505	7B3O	Bertoglio et al. [15]
S2M11	SARS-CoV-2 RBD	Up protomer: 339, 342, 343, 345, 367, 368, 371–374, 436, 440, 441, 444, 339, 342, 343, 345, 367, 368, 371–374, 436, 440, 441, 444 Down protomer: 446, 447, 449, 452, 455, 456, 483–490, 492–494, 496, 498	7K43	Tortorici et al. [16]

Appendix D
Vita

Yui Tik (Andrew) Pang was born in Hong Kong on July 10, 1993. He received the degree of Bachelor of Science in Physics and Master of Philosophy in Physics from the Chinese University of Hong Kong in 2012 and 2015, respectively. Andrew's primary research interest lies in the field of Biophysics. However, his academic curiosity extends beyond this, as he also possesses a robust interest in Computer Science and Parallel Computing. His diverse interests allow him to approach his research from a unique interdisciplinary perspective, combining the principles of Physics and Computer Science to explore and solve complex biological problems. His latest endeavor involves integrating deep learning techniques with molecular mechanics calculations to predict novel protein conformations.

List of Publications

1. Chiang, Y.; **Pang, Y. T.**; Wang, Y. The role of intramolecular nonbonded interaction and angle sampling in single-step free energy perturbation. *J Chem Phys* **2016**, *145* (23), 234109. AIP Publishing.
2. **Pang, Y. T.**; Miao, Y.; Wang, Y.; McCammon, J. A. Gaussian accelerated molecular dynamics in NAMD. *J Chem Theory Comput* **2017**, *13* (1), 9–19. American Chemical Society.
3. **Pang, Y. T.**; Ge, Z.; Zhang, B.; Xiu, P.; Li, Q.; Wang, Y. Pore formation induced by nanoparticles binding to a lipid membrane. *Nanoscale* **2020**, *12* (14), 7902–7913. Royal Society of Chemistry.
4. Ryoo, D.; Rydmark, M. O.; **Pang, Y. T.**; Lundquist, K. P.; Linke, D.; Gumbart, J. C. BamA is required for autotransporter secretion. *Biochim Biophys Acta Gen Subj* **2020**, *1864* (7), 129581. Elsevier.
5. Nassiri, T. Z.; Su, X.; Austin, R.; Choudhury, S.; Li, W.; **Pang, Y. T.**; Gumbart, J. C.; Torres, M. P. Combinatorial phosphorylation modulates the structure

Y. T. (Andrew) Pang, *Biomolecules' Conformational Changes Studied by Simulations and Enhanced Sampling*, Springer Theses,
https://doi.org/10.1007/978-3-031-70602-8

and function of the G protein γ subunit in yeast. *Sci Signal* **2021**, *14* (688), eabd2464. American Association for the Advancement of Science.

6. **Pang, Y. T.**; Pavlova, A.; Tajkhorshid, E.; Gumbart, J. C. Parameterization of a drug molecule with a halogen σ-hole particle using ffTK: Implementation, testing, and comparison. *J Chem Phys* **2020**, *153* (16), 164104. AIP Publishing.
7. Pavlova, A.; Lynch, D. L.; Daidone, I.; Zanetti-Polzi, L.; Smith, M. D.; Chipot, C.; Kneller, D. W.; Kovalevsky, A.; Coates, L.; Golosov, A. A.; Dickson, C. J.; Velez-Vega, C.; Duca, J. S.; Vermaas, J. V.; **Pang, Y. T.**; Acharya, A.; Parks, J. M.; Smith, J. C.; Gumbart, J. C. Inhibitor binding influences the protonation states of histidines in SARS-CoV-2 main protease. *Chem Sci* **2021**, *12* (4), 1513–1527. Royal Society of Chemistry.
8. Acharya, A.; Lynch, D. L.; Pavlova, A.; **Pang, Y. T.**; Gumbart, J. C. ACE2 glycans preferentially interact with SARS-CoV-2 over SARS-CoV. *Chem Commun (Camb)* **2021**, *57* (48), 5949–5952. Royal Society of Chemistry.
9. Pavlova, A.; Zhang, Z.; Acharya, A.; Lynch, D. L.; **Pang, Y. T.**; Mou, Z.; Parks, J. M.; Chipot, C.; Gumbart, J. C. Machine learning reveals the critical interactions for SARS-CoV-2 spike protein binding to ACE2. *J Phys Chem Lett* **2021**, *12* (23), 5494–5502. American Chemical Society.
10. Wu, R.; Bakelar, J. W.; Lundquist, K.; Zhang, Z.; Kuo, K. M.; Ryoo, D.; **Pang, Y. T.**; Sun, C.; White, T.; Klose, T.; Jiang, W.; Gumbart, J. C.; Noinaj, N. Plasticity within the barrel domain of BamA mediates a hybrid-barrel mechanism by BAM. *Nat Commun* **2021**, *12* (1), 7131. Nature Publishing Group UK London.
11. **Pang, Y. T.**; Acharya, A.; Lynch, D. L.; Pavlova, A.; Gumbart, J. C. SARS-CoV-2 spike opening dynamics and energetics reveal the individual roles of glycans and their collective impact. *Commun Biol* **2022**, *5* (1), 1170. Nature Publishing Group UK London.
12. **Pang, Y. T.**; Hazel, A. J.; Gumbart, J. C. Uncovering the folding mechanism of pertactin: A comparative study of isolated and vectorial folding. *Biophys J* **2023**, *122* (14), 2988–2995. Elsevier.
13. Su, X.; **Pang, Y. T.**; Li, W.; Gumbart, J.; Kelley, J.; Torres, M. N-terminal intrinsic disorder is an ancestral feature of Gγ subunits that influences the balance between different G$\beta\gamma$ signaling axes in yeast. *J Biol Chem* **2023**, *299* (8), 104947. Elsevier.
14. **Pang, Y. T.**; Yang, L.; Gumbart, J. C. From simple to complex: Reconstructing all-atom structures from coarse-grained models using cg2all. *Structure* **2024**, *32* (1), 5–7. Elsevier.
15. Zeng, Y.; Pavlova, A; Nelson, P. M.; Glick, Z. L.; Yang, L.; **Pang, Y. T.**; Spivak, M.; Licari, G; Tajkhorshid, E; Sherrill, C. D.; Gumbart, J. C. Broadening access to small-molecule parameterization with the force field toolkit. *J Chem Phys* **2024**, *160* (24), 242501. AIP Publishing.
16. Satalkar, V.; Degaga, G. D.; Li, W.;, **Pang, Y. T.**; McShan, A. C.; Gumbart, J. C.; Mitchell, J. C.; Torres, M. P. Generative β-hairpin design using a residue-based physicochemical property landscape. *Biophys J* **2024**, In press. Elsevier.

References

1. B. Ensing, A. Laio, M. Parrinello, M.L. Klein, A recipe for the computation of the free energy barrier and the lowest free energy path of concerted reactions. J. Phys. Chem. B **109**(14), 6676–6687 (2005)
2. M. Moradi, V. Babin, C. Roland, T.A. Darden, C. Sagui, Conformations and free energy landscapes of polyproline peptides. Proc. Natl. Acad. Sci. U. S. A. **106**(49), 20746–20751 (2009)
3. P. Virtanen et al., SciPy 1.0: fundamental algorithms for scientific computing in Python. Nat. Methods **17**, 261–272 (2020)
4. M.V. Smoluchowski, Über Brownsche Molekularbewegung unter Einwirkung äußerer Kräfte und deren Zusammenhang mit der verallgemeinerten Diffusionsgleichung. Ann. Phys. **353**(24), 1103–1112 (1916)
5. A. Singharoy, C. Chipot, M. Moradi, K. Schulten, Chemomechanical coupling in hexameric protein-protein interfaces harnesses energy within V-type ATPases. J. Am. Chem. Soc. **139**(1), 293–310 (2017)
6. K. Gaalswyk, E. Awoonor-Williams, C.N. Rowley, Generalized Langevin methods for calculating transmembrane diffusivity. J. Chem. Theory Comput. **12**(11), 5609–5619 (2016)
7. A. Fakharzadeh, M. Moradi, Effective Riemannian diffusion model for conformational dynamics of biomolecular systems. J. Phys. Chem. Lett. **7**(24), 4980–4987 (2016)
8. J. Lan et al., Structure of the SARS-CoV-2 spike receptor-binding domain bound to the ACE2 receptor. Nature **581**, 215–220 (2020)
9. D. Cremer, J.A. Pople, General definition of ring puckering coordinates. J. Am. Chem. Soc. **97**(6), 1354–1358 (1975)
10. L. Liu et al., Potent neutralizing antibodies against multiple epitopes on SARS-CoV-2 spike. Nature **584**, 450–456 (2020)
11. X. Chi et al., A neutralizing human antibody binds to the N-terminal domain of the Spike protein of SARS-CoV-2. Science **369**, 650–655 (2020)
12. S. Du et al., Structurally resolved SARS-CoV-2 antibody shows high efficacy in severely infected hamsters and provides a potent cocktail pairing strategy. Cell **183**, 1013–1023 (2020)
13. D. Pinto et al., Cross-neutralization of SARS-CoV-2 by a human monoclonal SARS-CoV antibody. Nature **583**, 290–295 (2020)
14. M. Yuan et al., A highly conserved cryptic epitope in the receptor binding domains of SARS-CoV-2 and SARS-CoV. Science **368**(6491), 630–633 (2020)
15. F. Bertoglio et al., SARS-CoV-2 neutralizing human recombinant antibodies selected from pre-pandemic healthy donors binding at RBD-ACE2 interface. Nat. Commun. **12**, 1577 (2021)
16. M.A. Tortorici et al., Ultrapotent human antibodies protect against SARS-CoV-2 challenge via multiple mechanisms. Science **370**, 950–957 (2020)

SPRINGER NATURE

GPSR Compliance

The European Union's (EU) General Product Safety Regulation (GPSR) is a set of rules that requires consumer products to be safe and our obligations to ensure this.

If you have any concerns about our products, you can contact us on ProductSafety@springernature.com

In case Publisher is established outside the EU, the EU authorized representative is:

Springer Nature Customer Service Center GmbH
Europaplatz 3
69115 Heidelberg, Germany

The manufacturer's authorised representative in the EU is Springer Nature Customer Service Centre GmbH, Europaplatz 3, 69115 Heidelberg, Germany. If you have any concerns regarding our products, please contact ProductSafety@springernature.com

Printed and bound by CPI Group (UK) Ltd, Croydon, CR0 4YY
30/09/2026
02225577-0001